Foundations of Radiation Theory and Quantum Electrodynamics

Foundations of Radiation Theory and Quantum Electrodynamics

Edited by
A. O. Barut

University of Colorado
Boulder, Colorado

Springer Science+Business Media, LLC

Library of Congress Cataloging in Publication Data

Main entry under title:

Foundations of radiation theory and quantum electrodynamics.

"Much of the material was presented at a recent symposium held at the University of Colorado in Boulder."
Includes index.
1. Quantum electrodynamics. 2. Radiation. I. Barut, Asim Orhan, 1926-
QC680.F68   537.6   79-25715
ISBN 978-1-4757-0673-4   ISBN 978-1-4757-0671-0 (eBook)
DOI 10.1007/978-1-4757-0671-0

Contributors

A. O. Barut, Department of Physics, University of Colorado, Boulder, Colorado

Iwo Bialynicki-Birula, Institute of Theoretical Physics, Warsaw University, Poland, and Department of Physics, University of Pittsburgh, Pittsburgh, Pennsylvania

Timothy H. Boyer, Department of Physics, City College of the City University of New York, New York, New York

Luiz Davidovich, Instituto de Física, Pontifícia Universidade Católica do Rio de Janeiro, Brazil

J. H. Eberly, Joint Institute for Laboratory Astrophysics, University of Colorado and National Bureau of Standards, Boulder, Colorado

H. Grotch, Department of Physics, Pennsylvania State University, University Park, Pennsylvania

E. T. Jaynes, Department of Physics, Washington University, St. Louis, Missouri

E. Kazes, Department of Physics, Pennsylvania State University,, University Park, Pennsylvania

Darryl Leiter, Code 660, NASA/Goddard Space Flight Center, Greenbelt, Maryland

Peter W. Milonni, Perkin-Elmer Corporation, Wilton, Connecticut

Kimball A. Milton, Department of Physics, University of California, Los Angeles, California

Michael Martin Nieto, Theoretical Division, Los Alamos Scientific Laboratory, University of California, Los Alamos, New Mexico

H. M. Nussenzveig, Instituto de Física, Universidade de São Paulo, São Paulo, Brazil

E. A. Power, Department of Mathematics, University College London, London, England

F. Rohrlich, Department of Physics, Syracuse University, Syracuse, New York

Marlan O. Scully, Department of Physics and Optical Sciences Center, University of Arizona, Tucson, Arizona

D. H. Sharp, Theoretical Division, Los Alamos Scientific Laboratory, University of California, Los Alamos, New Mexico

L. M. Simmons, Jr., Theoretical Division, Los Alamos Scientific Laboratory, University of California, Los Alamos, New Mexico

T. Thirunamachandran, Department of Chemistry, University College London, London, England

K. Wódkiewicz, Institute of Theoretical Physics, Warsaw University, Warsaw, Poland

Preface

Quantum theory originated in the study of the properties of electromagnetic radiation and of the interactions of electromagnetic radiation with matter. Yet these problems still dominate much of the work on the frontiers of research. New approaches to radiation theory help us to gain a deeper understanding of the underlying physical phenomena for which we still do not have a complete and closed theory and which may still bring new surprises. At the same time the scope and applications of radiation phenomena have increased enormously; whole new fields, such as quantum optics and laser physics, have developed, and the precision measurements possible with electromagnetic radiation abound in all areas of physics.

We felt therefore that the time has come to look again at the foundations of radiation theory and to determine where we stand. We attempt here to collect in one volume various approaches to the problems of radiation theory, quantum electrodynamics, and electron theory. We hope that these contributions will help to elucidate the current status and problems of the field, provide reviews and syntheses, and serve as source material for references to the literature.

Much of the material was presented at the recent symposium on the Foundations of Radiation Theory and Quantum Electrodynamics held at the University of Colorado in Boulder. The authors have, however, greatly expanded their contributions in such a way that this volume will be, we hope, of more permanent value. I should like to thank here our contributors for their efforts and enthusiasms.

Boulder   A. O. Barut

Contents

Classical and Quantum Theories of Radiation

Peter W. Milonni

> It is obvious, however, that whatever side we
> take concerning the nature of light, many, indeed
> almost all the circumstances concerning it, are
> incomprehensible, and beyond the reach of
> human understanding—*Encyclopaedia Britannica*
> article on "Light," 1792

1. Introduction

Quantum theory developed from the study of the interaction of light
and matter. Around the beginning of this century, experimental physicists
reported phenomena that demonstrated the inadequacy of classical physics. Quantum features of the electromagnetic field were first postulated
by Planck in 1900 in order to account for the spectrum of blackbody
radiation observed by Lummer and Pringsheim. In order to account for
the discrete sequence of wavelengths in the spectrum of atomic hydrogen,
Bohr in 1913 postulated that the electron could move only in certain
stationary orbits; Planck's constant appeared in his theory as the fundamental "unit of action." We are all familiar with these and other
inspired guesses which led finally to the birth of quantum mechanics, as
we know it, in the years 1925–26. Dirac's paper on the quantum theory
of the electromagnetic field appeared in 1927,[1] and by 1930 there was
little doubt that the quantum theory of light and matter was vastly su-
perior to classical theory. Born was so impressed by the theory that at

Peter W. Milonni • Perkin-Elmer Corporation, Wilton, Connecticut. This work is ded-
icated to the memory of Renate Wiener Chasman.

the Fifth Solvay Conference in 1927 he remarked, "We consider that quantum mechanics is a complete theory, and that its fundamental hypotheses, both physical and mathematical, are not susceptible to further modification."

In this introductory lecture I shall review some developments, old and new, in electrodynamics. The choice of topics was guided by our intention here to compare the classical and quantum theories of radiation. My approach is to emphasize physical concepts as much as possible without detailed calculations.

2. The Origin of Commutation Relations

I shall begin by reviewing how commutation relations made their way into quantum theory. This bit of prehistory will not be strictly accurate. An historically accurate account would require a rather detailed discussion, including Heisenberg's "quantum-theoretical reinterpretation" of classical dispersion theory.[2] My intention is to review an *example* of the type of reasoning that led to the formalism of quantum mechanics.

Ladenburg[3] generalized the Maxwell–Sellmeier dispersion formula by replacing the classical number of dispersion electrons by a quantity obtained from an examination of Einstein's derivation[4] of Planck's blackbody spectrum. The Ladenburg formula for the refractive index near the atomic (circular) transition frequency ω_{ki} is

$$n^2(\omega) = 1 + \frac{2\pi c^3 N_i A_{ki}}{\omega_{ki}^2(\omega_{ki}^2 - \omega^2)} \tag{1}$$

where N_i is the number density of atoms in the lower state i of the transition $k \rightarrow i$ and A_{ki} is the Einstein coefficient of spontaneous emission for the transition. If we add the contributions of all possible excited states that make (allowed) transitions to state i, and demand that the polarizability

$$\alpha_i(\omega) = \tfrac{1}{2}c^3 \sum_k \frac{A_{ki}}{\omega_{ki}^2(\omega_{ki}^2 - \omega^2)} \tag{2}$$

should reduce to the Thomson free-electron formula

$$\alpha(\omega) = -e^2/m\omega^2 \tag{3}$$

in the high-frequency x-ray limit, then we obtain the Thomas–Reiche–Kuhn sum rule,

$$\frac{2m}{3\hbar} \sum_k |\mathbf{r}_{ki}|^2 \omega_{ki} = 1 \tag{4}$$

We have used the fact that $A_{ki} = 4e^2 |\mathbf{r}_{ki}|^2 \omega_{ki}^3 / 3\hbar c^3$, which was derived by Kramers and Heisenberg from the Correspondence Principle. We have essentially reproduced the line of reasoning used by Kuhn[5] in his derivation of the sum rule. The significance of $\mathbf{r}_{ki}$ as a "matrix element" connecting *two* atomic stationary states (or Bohr orbits) stems from the work of Heisenberg[2] and Born and Jordan.[6] In his attempt to deal as far as possible with observable entities, Heisenberg concentrated his attention on atomic transitions rather than term values (energy levels). Sets of numbers, each involving two states, were conveniently arranged in rows and columns. Born had attended lectures on noncommutative algebras by Rosanes and was evidently the first physicist to recognize the matrix character of Heisenberg's transition elements.

Equation (4) is just a statement of the commutation rule, $[x,p] = i\hbar$. This commutation rule first appeared in the paper of Born and Jordan,[6] who invoked the "quantization condition" (4) in postulating the commutation rule.

It is worth mentioning Kramers' modification of the Ladenburg dispersion formula (1) to include the possibility of populating excited states. Kramers[7] replaced (1) with

$$n^2(\omega) = 1 + \frac{2\pi c^3 A_{ki}(N_i - N_k)}{\omega_{ki}^2(\omega_{ki}^2 - \omega^2)} \qquad (5)$$

where N_k is the number of atoms per unit volume in the excited state k. Kramers arrived at his "negative oscillator" term in the dispersion formula by considering Einstein's derivation of the Planck formula: If the rate at which radiation is absorbed depends upon the *difference* between the population densities, then so too should the dispersion.

Ladenburg and Kopfermann[8] in 1928 reported their experimental verification of Kramers' negative oscillators, i.e., of stimulated emission. They passed one beam of a differential refractometer through a discharge tube filled with neon, and the other beam passed through an unexcited but otherwise identical tube. Thus the difference in the refractive indices of the tubes could be measured. Figure 1 shows their result, plotted schematically against the current through the excited discharge, for the refractive index near the 5882 Å transition of Ne. For low currents the index in the excited tube increases with increasing current, since the population N_i of the lower level of the transition is increasing. But as the current is increased the population difference $N_i - N_k$ increases less rapidly and eventually decreases.

Obviously the result might be interpreted not as an increase in N_k compared with N_i, but simply as a decrease in N_i at the higher currents. In order to decide between these two alternatives, Ladenburg and Kopfermann measured the index near several spectral lines, all of which had

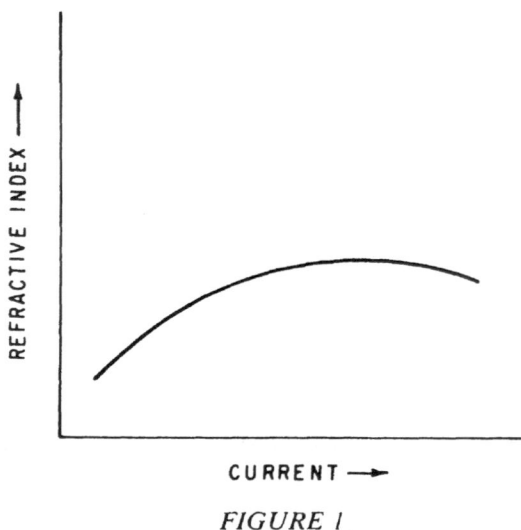

FIGURE 1

the same lower level *i*. They found that the dispersion curve of Figure 1 was different for the various lines, thus ruling out the alternative explanation.

This was the first direct experimental verification of stimulated emission. It showed, moreover, that the effect was "coherent," since it contributed to the dispersion. This important work of Ladenburg and Kopfermann seems now to have been forgotten. Of course, the concept of stimulated emission itself is another of those ideas which first appeared in Einstein's paper on the blackbody spectrum. Without the stimulated emission term Einstein's approach would have led to Wien's law, not Planck's.

The sign of the population difference $N_i - N_k$ [cf. equation (5)] determines the relative phase between the incoming field and the field radiated by the induced atomic dipole. Whether the incoming field is amplified by stimulated emission or attenuated by absorption depends upon this phasing and the consequent interference of the incoming and scattered fields.

3. Spontaneous Emission

Although Dirac's 1927 paper is usually considered to be the first work on field quantization, the radiation field was treated quantum mechanically in the Born–Heisenberg–Jordan paper of 1926.[9] These authors recognized that the field could be treated as a set of uncoupled harmonic oscillators. The quantization of the harmonic oscillator thus formalized Einstein's conception of field quanta, for the theory *predicted* the energy spectrum (photons) of the radiation field.

Dirac considered the old problem of calculating the rate of spontaneous emission. The answer was already known, as we have mentioned, from the Correspondence Principle; the problem was to show that the result was a natural consequence of the quantum theory. The quantum theory could not be considered a truly successful theory if it could not deal with this process in which a particle (photon) is actually created. According to Weinberg, "Dirac's successful treatment of the spontaneous emission of radiation confirmed the universal character of quantum mechanics."[10]

There are certain features of the problem that are obscured by the usual perturbation-theoretic calculation of the transition rate. I shall therefore outline a more recent approach to the problem.[11]

Consider an excited atom in free space, with no other sources of radiation to act upon it. For simplicity, but without much loss of generality for our purposes here, I shall consider the two-state model of a one-electron atom. The Hamiltonian for the atom–field system with "minimal coupling" is

$$H = H_A + H_F - (e/mc)\mathbf{A} \cdot \mathbf{p} \qquad (6)$$

where H_A and H_F are the Hamiltonian operators for the atom and field, respectively, and $\mathbf{A}$ is the (Coulomb-gauge) vector potential at the position of the atom (dipole approximation). The $\mathbf{A}^2$ term may be omitted for our purposes. The Hamiltonian (6) may be written as[11]

$$H = \tfrac{1}{2}\hbar\omega_0\sigma_z + \sum_{k,\lambda} \hbar\omega_k a_{k\lambda}^\dagger a_{k\lambda} + i\hbar \sum_{k,\lambda} C_{k\lambda}(a_{k\lambda} + a_{k\lambda}^\dagger)(\sigma - \sigma^\dagger) \qquad (7)$$

Here $a_{k\lambda}$ is the photon annihilation operator for the plane-wave field mode with wave vector $\mathbf{k}$ and polarization index λ, and $\omega_k = kc$. The σ's are Pauli spin-half operators, in the conventional notation, and ω_0 is the (circular) transition frequency between the two atomic states under consideration. The coupling constant is

$$C_{k\lambda} = (2\pi/\hbar\omega_k V)^{1/2}\omega_0\boldsymbol{\mu} \cdot \hat{\mathbf{e}}_{k\lambda} \qquad (8)$$

where V is the quantization volume (eventually allowed to be infinite) and $\boldsymbol{\mu}$ is the electric-dipole matrix element between the two atomic states, taken to be real. Finally, $\hat{\mathbf{e}}_{k\lambda}$ is a polarization unit vector such that $\hat{\mathbf{e}}_{k\lambda} \cdot \hat{\mathbf{e}}_{k\lambda'} = \delta_{\lambda\lambda'}$, $\mathbf{k} \cdot \hat{\mathbf{e}}_{k\lambda} = 0$, $\lambda = 1, 2$.

The field acting on the atom may be divided conceptually into two parts. One is the field from the atom, acting back on the atom; this is the radiation reaction field. The other is the zero-point or vacuum field. We have tacitly discarded the zero-point energy $\tfrac{1}{2}\sum_{k,\lambda}\hbar\omega_k$ from the Hamiltonian (7), but we have not by any means eliminated the vacuum field from the problem. It will reappear as the homogeneous (source-free) solution of the operator Maxwell equation for the vector potential or electric field, i.e., in the Heisenberg equation of motion for the field. Its expectation value is zero, but we certainly cannot set the vacuum field *operator* equal to zero on this account.

Which of these two fields is physically "responsible" for the atom dropping to the lower level and emitting radiation? The well-known folk theorem states that the spontaneous emission is caused by the vacuum field. This interpretation suggests that the fluctuating vacuum field somehow "shakes loose" the stored atomic energy. But classically we attribute the damping of an oscillating dipole to the force of radiation reaction. The traditional ways of calculating the spontaneous emission rate do not shed any light on this question of interpretation. The Heisenberg-picture approach, on the other hand, is ideally suited to address questions of interpretation.

In order to solve Heisenberg equations of motion for the atomic and field operators we must know the various commutation relations. The commutation relations among atomic operators are known, as are the field commutation relations. It remains therefore to specify commutation relations between atomic and field operators. We shall require that *equal-time* atomic and field operators commute, i.e.,

$$[\sigma(t), a_{k\lambda}(t)] = 0, \text{ etc.} \tag{9}$$

This is consistent with the assumption that the interaction is switched on at some time $t = 0$ at which the atom and field are uncoupled and their respective operators act in different Hilbert spaces. The state vector at $t = 0$ is taken to be

$$|\psi\rangle = |+\rangle \otimes |\{0\}\rangle \tag{10}$$

i.e., the atom is initially in the upper state $|+\rangle$ and the field state is the vacuum state of no "photons."

Suppose we choose a *normal ordering*, in which photon annihilation (creation) operators are placed to the extreme right (left) in all operator products. Remember that the assumption that atom and field operators commute means that the ordering of the σ's and a's, and their adjoints, is inconsequential (except, as we shall see, for matters of interpretation). The Heisenberg equations of motion for the operators σ, σ_z, and $a_{k\lambda}$ are

$$\dot{\sigma}(t) = -i\omega_0 \sigma(t) + \sum_{k,\lambda} C_{k\lambda}[\sigma_z(t)a_{k\lambda}(t) + a_{k\lambda}^{\dagger}(t)\sigma_z(t)] \tag{11}$$

$$\dot{\sigma}_z(t) = -2 \sum_{k,\lambda} C_{k\lambda}[\sigma(t) + \sigma^{\dagger}(t)]a_{k\lambda}(t)$$

$$-2 \sum_{k,\lambda} C_{k\lambda}a_{k\lambda}^{\dagger}(t)[\sigma(t) + \sigma^{\dagger}(t)] \tag{12}$$

$$\dot{a}_{k\lambda}(t) = -i\omega_k a_{k\lambda}(t) + C_{k\lambda}[\sigma(t) - \sigma^{\dagger}(t)] \tag{13}$$

when the field operators are normally ordered. We now find an approximate expression for the operator

$$\sum_{k,\lambda} C_{k\lambda} a_{k\lambda}(t) = (\omega_0/\hbar c)\boldsymbol{\mu} \cdot \mathbf{A}^{(+)}(0, t) \tag{14}$$

and its Hermitian conjugate and substitute the result into equations (11) and (12). $\mathbf{A}^{(+)}(0,t)$ is the operator representing the positive-frequency part of the vector potential at the position of the atom. Within an "adiabatic approximation" that is the Heisenberg-picture equivalent of the Weisskopf–Wigner approximation[11] it is found that

$$(\omega_0/\hbar c)\boldsymbol{\mu} \cdot \mathbf{A}^{(+)}(0, t) \cong (\omega_0/\hbar c)\boldsymbol{\mu} \cdot \mathbf{A}_0^{(+)}(0, t) + \beta\sigma(t)$$
$$- i[\Delta^{(-)}\sigma(t) - \Delta^{(+)}\sigma^\dagger(t)] \tag{15}$$

$$(\omega_0/\hbar c)\boldsymbol{\mu} \cdot \mathbf{A}^{(-)}(0, t) \cong (\omega_0/\hbar c)\boldsymbol{\mu} \cdot \mathbf{A}_0^{(-)}(0, t) + \beta\sigma^\dagger(t)$$
$$- i[\Delta^{(+)}\sigma(t) - \Delta^{(-)}\sigma^\dagger(t)] \tag{16}$$

where $\beta = 2\mu^2\omega_0^3/3\hbar c^3$ is half the Einstein A-coefficient for spontaneous emission and

$$\Delta^{(\pm)} = \frac{\beta}{\pi\omega_0} P \int_0^\infty d\omega \frac{\omega}{\omega \pm \omega_0} \tag{17}$$

$\mathbf{A}_0^{(+)}(0, t)$ is the positive-frequency part of the free or vacuum solution for the vector potential:

$$\mathbf{A}_0^{(+)}(\mathbf{r}, t) = \sum_{k,\lambda} \left(\frac{2\pi\hbar c^2}{\omega_k V}\right)^{1/2} a_{k\lambda}(0)\exp[-i(\omega_k t - \mathbf{k}\cdot\mathbf{r})]\hat{e}_{k\lambda} \tag{18}$$

The advantage of normal ordering is realized when we substitute (15) and (16) into (11) and (12) and take expectation values over the initial state (10): The free-field operators do not contribute explicitly to the expectation values. In the "rotating-wave approximation" of neglecting $\sigma^\dagger$ terms in the equation of motion for σ, we find from equation (11) that

$$\langle \dot\sigma(t) \rangle = -i\omega_0\langle\sigma(t)\rangle + \beta\langle\sigma_z(t)\sigma(t)\rangle - i[\Delta^{(-)}\langle\sigma_z(t)\sigma(t)\rangle$$
$$+ \Delta^{(+)}\langle\sigma(t)\sigma_z(t)\rangle] \tag{19}$$

Now $\sigma_z(t)\sigma(t) = -\sigma(t)\sigma_z(t) = -\sigma(t)$, so that

$$\langle\dot\sigma(t)\rangle = -i\omega_0\langle\sigma(t)\rangle - \beta\langle\sigma(t)\rangle + i(\Delta^{(-)} - \Delta^{(+)})\langle\sigma(t)\rangle$$
$$= -i[\omega_0 - (\Delta^{(-)} - \Delta^{(+)}) - i\beta]\langle\sigma(t)\rangle \tag{20}$$

From this result we may identify the frequency shift

$$\delta = -(\Delta^{(-)} - \Delta^{(+)}) \tag{21}$$

and linewidth β. Moreover, since normal ordering has been used to eliminate explicit contributions from the vacuum field $\mathbf{A}_0(\mathbf{r}, t)$, we see that the line shift and width may be attributed to the radiation reaction field—the solution of the inhomogeneous Maxwell equation for the vector potential.

If we write $\hbar\omega_0 = E_2 - E_1$, we see that $\hbar(\omega_0 + \delta) = (E_2 - \hbar\Delta^{(-)}) - (E_1 - \hbar\Delta^{(+)})$. In fact, it can be shown explicitly that $-\hbar\Delta^{(-)}$ is the level shift of the upper level 2, and $-\hbar\Delta^{(+)}$ is the level shift of the lower level 1.[11] Level shifts are superfluous in the Heisenberg-picture approach; the *frequency shifts* come out naturally. The level shifts are just those to which Bethe applied mass renormalization to obtain a (finite) number for the Lamb shift.[12] A multilevel calculation of the radiative shift in level j yields

$$\Delta E_j = - \frac{2}{3\pi c^3} \sum_m \omega_{jm}^2 |\boldsymbol{\mu}_{jm}|^2 P \int_0^\infty d\omega \frac{\omega}{\omega - \omega_{jm}} \tag{22}$$

where $\hbar\omega_{jm} = E_j - E_m$ and $\boldsymbol{\mu}_{jm}$ is the electric-dipole matrix element between states j and m. In terms of the momentum matrix element $\mathbf{p}_{jm}$ we may write (22) as

$$\Delta E_j = \frac{2}{3\pi} \left(\frac{e^2}{\hbar c}\right)\left(\frac{1}{mc}\right)^2 \sum_m |\mathbf{p}_{jm}|^2 P \int_0^\infty dE \frac{E}{E_j - E_m - E} \tag{23}$$

Bethe[12] subtracted from ΔE_j the *free*-electron energy $\Delta E_j^{(F)}$ [obtained by setting $E_j - E_m = 0$ in (23)],

$$\Delta E_j^{(F)} = - \frac{2}{3\pi}\left(\frac{e^2}{\hbar c}\right)\left(\frac{1}{mc}\right)^2 \sum_m |\mathbf{p}_{jm}|^2 \int_0^\infty dE \tag{24}$$

to obtain the observable shift

$$\Delta E_j^{(obs)} = \frac{2}{3\pi}\left(\frac{e^2}{\hbar c}\right)\left(\frac{1}{mc}\right)^2 \sum_m (E_j - E_m)|\mathbf{p}_{jm}|^2 \int_0^{E_{max}} \frac{dE}{E_j - E_m - E}$$

$$= - \frac{2}{3\pi}\left(\frac{e^2}{\hbar c}\right)\left(\frac{1}{mc}\right)^2 \sum_m (E_j - E_m)|\mathbf{p}_{jm}|^2 \log\frac{mc^2}{|E_j - E_m|} \tag{25}$$

where the cutoff energy E_{max} (taken by Bethe to be mc^2) is introduced on the grounds that the nonrelativistic theory is only applicable for photon energies $E \ll mc^2$; thus the Lamb shift was assumed *a priori* to be predominantly a nonrelativistic effect. Since $mc^2 \gg |E_j - E_m|$, Bethe took the logarithm in (25) to be constant (independent of state m) as a first approximation. Then, since

$$\sum_m (E_j - E_m)|\mathbf{p}_{jm}|^2 = -\tfrac{1}{2}\hbar^2 \int d^3r \, \nabla^2 V |\psi(\mathbf{r})|^2$$

$$= -2\pi\hbar^2 e^2 |\psi_j(0)|^2 \tag{26}$$

it follows that

$$\Delta E_j^{(\text{obs})} \cong \frac{4}{3}\left(\frac{e^2}{hc}\right)\left(\frac{e\hbar}{mc}\right)^2 |\psi_j(0)|^2 \log \frac{mc^2}{\langle E_j - E_m \rangle} \qquad (27)$$

where $\langle E_j - E_m \rangle$ is the *average* excitation energy of state j. Equation (27) indicates that only s states are shifted according to the Bethe theory. For the hydrogen $2s_{1/2}$ state Bethe calculated $\Delta E^{(\text{obs})}$ to be about 1040 Mc, in good agreement with the experimental result of Lamb and Retherford[13] for the $2s_{1/2}$-$2p_{1/2}$ splitting.

The solution of the hydrogen atom problem with the Dirac equation shows that states with the same n and j should be degenerate. Although a possible energy difference between the $2s_{1/2}$ and $2p_{1/2}$ states had been suggested in the 1930s by several workers, no definitive experiments were undertaken until Lamb and Retherford[13] utilized some techniques developed during wartime research in radar. Bethe is said to have done his calculations on the train to Ithaca following the Shelter Island conference. Subsequent relativistic calculations resulted in an entirely convergent value for the Lamb shift, as Bethe had guessed.[12] Mass renormalization nevertheless plays an important role in the relativistic calculation. It should also be noted that there was considerable confusion before the relativistic calculations could be considered as satisfactory.[14] A recent discussion of the significance of the renormalization idea in quantum field theory is given by Weinberg.[10]

At about the time that relativistic calculations were being performed, Welton[15] gave an elegant derivation of Bethe's result (27) in a paper on "Some Observable Effects of the Quantum-Mechanical Fluctuations of the Electromagnetic Field." Welton considered the motion of an otherwise free electron under the influence of the fluctuating vacuum field. It is easily found that such an electron has a mean-square fluctuation in position of

$$\langle (\Delta \mathbf{r})^2 \rangle = \frac{2e^2}{\pi \hbar c}\left(\frac{\hbar}{mc}\right)^2 \int_{k_{\min}}^{k_{\max}} \frac{dk}{k} \qquad (28)$$

As in Bethe's calculation a high-frequency cutoff is used ($k_{\max} = mc/\hbar$). The low-frequency cutoff is to be determined from the details of the electronic motion; a binding potential would suppress the low-frequency fluctuations. Now the average effective potential seen by the electron would be

$$\langle V(\mathbf{r} + \Delta \mathbf{r}) \rangle \cong V(\mathbf{r}) + \tfrac{1}{6}\langle (\Delta \mathbf{r})^2 \rangle \nabla^2 V(\mathbf{r}) \qquad (29)$$

so that for the Coulomb potential there is a correction to the potential given by

$$\Delta V(\mathbf{r}) = \frac{1}{6}\left(\frac{2e^2}{\pi\hbar c}\right)\left(\frac{\hbar}{mc}\right)^2 \log\left(\frac{mc^2}{\hbar c k_{\min}}\right) 4\pi e^2 \delta(\mathbf{r}) \qquad (30)$$

For a state described by the wave function $\psi(\mathbf{r})$ there is therefore a correction to the energy given by

$$\Delta E = \frac{4}{3}\left(\frac{e^2}{\hbar c}\right)\left(\frac{e\hbar}{mc}\right)^2 |\psi(0)|^2 \log\left(\frac{mc^2}{hck_{\min}}\right) \qquad (31)$$

which is Bethe's result (27) if $\hbar c k_{\min}$ is taken as Bethe's "average excitation energy." Welton's derivation makes it obvious that the energy shift should be *positive*, as the fluctuating influence of the zero-point field weakens the binding effect of the potential.

The Heisenberg-picture approach[11] implies that radiation reaction "causes" the level width and shift in spontaneous emission, and yet Welton's argument lends strong support to the idea that the vacuum field fluctuations are responsible for these radiative corrections. The resolution of this paradox is found when one considers field operator orderings other than the normal ordering used in the Heisenberg-picture approach discussed above.[11] If, for example, a completely symmetric ordering of positive- and negative-frequency field operators is used, the interpretation forced upon us is quite in line with Welton's. On the other hand, a more general type of ordering than normal or symmetrical attributes the radiative corrections neither to radiation reaction nor vacuum field fluctuations, but rather to a combination of the two.[11] Thus, each of the two interpretations is an oversimplification.[16]

4. Radiation Reaction

Since this lecture is intended to be an introduction to those which follow, it may be worthwhile to consider in more detail the issue of radiation reaction. If we solve formally for the operator $a_{k\lambda}(t)$ in equation (13) and use the result in the expression

$$\mathbf{A}(0, t) = \sum_{\mathbf{k},\lambda}\left(\frac{2\pi\hbar c^2}{\omega_k V}\right)^{1/2} [a_{k\lambda}(t) + a^\dagger_{k\lambda}(t)]\hat{\mathbf{e}}_{k\lambda} \qquad (32)$$

we find that we can express the solution of the inhomogeneous equation as

$$\mathbf{A}_{\mathrm{RR}}(t) = \frac{4e}{3\pi mc^2}[Kc\mathbf{p}(t) - \frac{\pi}{2}\dot{\mathbf{p}}(t)] \qquad (33)$$

where again the cutoff $K = mc/\hbar$ is used. From this we obtain the radiation reaction electric field *operator*

$$\mathbf{E}_{RR}(t) = \frac{2e}{3c^3} \dddot{\mathbf{r}}(t) - \frac{4eK}{3\pi c^2} \ddot{\mathbf{r}}(t) \tag{34}$$

The coefficient multiplying $\ddot{\mathbf{r}}(t)$ would be infinite without the cutoff, since a point charge has been assumed. A rigorous calculation for a finite charge distribution yields a coefficient inversely proportional to the radius of the distribution. Moreover, there are additional (higher-derivative) terms which vanish in the limit of a point charge.

If we substitute the field (34) into the equation of motion for an otherwise free electron, we have

$$\left(m_{\text{bare}} + \frac{4e^2K}{3\pi c^2} \right) \ddot{\mathbf{r}}(t) = m_{\text{obs}} \ddot{\mathbf{r}}(t) = \frac{2e^2}{3c^3} \dddot{\mathbf{r}}(t) \tag{35}$$

where the observed mass m_{obs} is some bare mass m_{bare} plus an "electromagnetic mass" $4e^2K/3\pi c^2$. If we take $K = m_{\text{obs}}c/\hbar$ ($E_{\text{max}} = m_{\text{obs}}c^2$), we find that the electromagnetic mass is very small compared with the observed electron mass ($4/3\pi$ times the fine-structure constant $\cong 3 \times 10^{-3}$), and we may therefore write

$$\frac{p^2}{2m_{\text{obs}}} \cong \frac{p^2}{2m_{\text{bare}}} - \frac{2}{3\pi} \left(\frac{e^2}{\hbar c} \right) \left(\frac{1}{m_{\text{obs}}c} \right)^2 E_{\text{max}}p^2 \tag{36}$$

We can now better appreciate Bethe's mass-renormalization approach to the Lamb shift. A contribution

$$\Delta E = -\frac{2}{3\pi} \left(\frac{e^2}{\hbar c} \right) \left(\frac{1}{mc} \right)^2 E_{\text{max}}p^2 \tag{37}$$

to the electron's energy arises from its interaction with its own radiation reaction field. But this contribution to the electron energy is already included when we write its energy as $p^2/2m_{\text{obs}}$ in the Schrödinger equation. When we calculate the electron's energy this contribution will reappear and so must be subtracted away, since it should only contribute once. When we note that

$$\sum_m |\mathbf{p}_{jm}|^2 = \langle j|\mathbf{p}^K|j \rangle \tag{38}$$

in equation (24), we see that the term $\Delta E_j^{(F)}$ that Bethe subtracted away is just the *expectation value* of ΔE [equation (37)] in state j.

These results should not be taken too seriously. A relativistic calculation shows that $\Delta E^{(F)}$ is only logarithmically divergent. But Bethe's calculation showed that one could obtain convergent answers from calculations giving apparently divergent expressions.

It is interesting to return to our two-state Heisenberg-picture calculation of spontaneous emission. If we examine equations (20), (15), and (16), we see that we can also write (20) as

$$\langle \dot{\sigma}(t) \rangle = -i\omega_0 \langle \sigma(t) \rangle - (\omega_0/\hbar c)\boldsymbol{\mu} \cdot [\langle \mathbf{A}_{RR}^{(+)}(t) \rangle - \langle \mathbf{A}_{RR}^{(-)}(t) \rangle] \quad (39)$$

within a rotating-wave approximation. Thus, it is not really the radiation reaction field as such which is inducing the line shift and width, but rather the *difference* between its positive- and negative-frequency parts. This is a purely quantum-mechanical feature. If $\mathbf{A}_{RR}(t) = \mathbf{A}_{RR}^{(+)}(t) + \mathbf{A}_{RR}^{(-)}(t)$ were inducing the line shift, we would obtain not equation (21) but rather

$$\delta' = -(\Delta^{(-)} + \Delta^{(+)}) \quad (40)$$

Notice that if in equation (19) we replace $\langle \sigma_z \sigma \rangle$ and $\langle \sigma \sigma_z \rangle$ by the decorrelated product $\langle \sigma_z \rangle \langle \sigma \rangle$, *as if the radiation reaction field were a classical (c-number) field*, we obtain a nonlinear equation in which δ' plays the role of a frequency-shift parameter. Making the same decorrelation in the equation for $\langle \sigma_z(t) \rangle$, we obtain the equations of the *neoclassical theory*,[17]

$$\dot{x} = \beta xz - \delta' yz \quad (41)$$

$$\dot{y} = \beta yz + \delta' xz \quad (42)$$

$$\dot{z} = -\beta(1 - z^2) \quad (43)$$

The neoclassical theory, like all semiclassical radiation theories, treats the electromagnetic field classically while treating matter quantum mechanically. The neoclassical theory has been the subject of intense debate during the last decade. It has been reviewed elsewhere[11] and does not require a lengthy discussion here. But it should be mentioned that the quantity δ', which plays the role of a frequency shift [albeit equations (41)–(43) predict a chirped frequency[17]], represents a striking difference from the Bethe mass-renormalization approach to the Lamb shift. We have

$$\delta' \cong -4\mu^2 \omega_0^2 K / 3\pi\hbar c^2 \quad (44)$$

Now consider a *classical* oscillator equation

$$e(\ddot{x} + \omega_0^2 x) = \frac{e^2}{m} E_{RR} \longrightarrow \frac{2\mu^2\omega_0}{\hbar}\left(\frac{2}{3c^3}\dddot{x} - \frac{4K}{3\pi c^2}\ddot{x}\right) \quad (45)$$

where we have used the Thomas–Reiche–Kuhn sum rule to replace e^2/m by $2\mu^2\omega_0/\hbar$. (This, of course, is not true for a two-state atom, since

$[x, p_x] = i\hbar$ does not apply in its finite-dimensional Hilbert space. We make this replacement in order not to introduce unimportant details. The replacement is a legitimate one if we consider an atom with negligible probability of being removed from its ground state.[11]) Now it can be shown that the ''adiabatic approximation'' and the approximations used to derive the neoclassical equations involve the replacement of $\ddot{x}$ with $-\omega_0^2 x$. (Thus these approaches to spontaneous emission do not predict ''runaway solutions''!) If we do this in equation (45), we obtain

$$\ddot{x}(t) + \omega_0^2 \left(1 - \frac{8\omega_0 \mu^2 K}{3\pi \hbar c^2} \right) x(t) = -2\beta \dot{x}(t) \tag{46}$$

which implies a frequency shift

$$\omega_0 \left(1 - \frac{8\omega_0 \mu^2 K}{3\pi \hbar c^2} \right)^{1/2} - \omega_0 \cong \delta' \tag{47}$$

Thus, the $\ddot{x}$ term in the radiation reaction field, which is usually associated with the electromagnetic mass, is translated in the (approximate) equations of the neoclassical theory into a frequency shift, i.e., the theory gives an entirely different interpretation to the term which in the usual mass-renormalization approach is subtracted away.

The problem of radiation reaction has still not been solved to everyone's satisfaction. Even when the infinite electromagnetic mass is cut off and we use equation (35), we incur the ''runaway'' problem, for in addition to the trivial solution $\ddot{x} = 0$, equation (35) has a solution

$$\dot{x} = (\text{const}) \exp(t/\lambda) \tag{48}$$

where $\lambda = 2e^2/3mc^3$. The runaway solution (48) cannot be ignored on physical grounds, for in an arbitrary force field it will not be true that $\dot{x}$, $\ddot{x}$, and higher derivatives vanish.

One way to avoid the runaway solution is to follow a suggestion of Dirac. For a charge unaffected by external forces except for an impulse at $t = 0$, for instance, the usual solution to the equation of motion

$$\ddot{x}(t) = A\delta(t) + \lambda \dddot{x}(t) \tag{49}$$

would be

$$\dot{x}(t) = \begin{cases} 0 & t < 0 \\ A[1 - \exp(t/\lambda)] & t > 0 \end{cases} \tag{50}$$

Dirac suggested instead the solution

$$\dot{x}(t) = \begin{cases} A \exp(t/\lambda) & t < 0 \\ A & t > 0 \end{cases} \tag{51}$$

This solution is bounded, but the price paid is a violation of causality, since the electron motion *anticipates* the impulse force at $t = 0$.

The Feynman–Wheeler *absorber theory of radiation*[18] eliminates these difficulties, but it has not won widespread support.[19] Dirac had proposed the elimination of the divergent self-energy by taking not the usual retarded solution of the Maxwell equation, but rather half the difference of retarded and advanced fields. This eliminates the $\dddot{x}$ term in the radiation reaction field, and the first nonvanishing term is then the $\ddot{x}$ term which is needed for conservation of energy. Feynman and Wheeler[18] proposed an action-at-a-distance theory in which a point charge does not interact with itself, but only with other charges, and that it does this via half the retarded and half the advanced fields of the other charges. The radiative damping term ($\sim \dddot{x}$) arises because of the absorption of the outgoing field of a charge by the rest of the universe. But the runaway solution is prevented, since the resulting unbounded field could not be completely absorbed.

The infinite self-interaction problem is not "solved" in the relativistic theory, although the order of the infinity is reduced. Moreover, the theory introduces another infinity (vacuum polarization) that is nonclassical. Modern field theory extracts finite and apparently correct results through elaborate renormalization procedures.[10]

5. The Vacuum Field

In our discussion of spontaneous emission we found it necessary to consider not only the field of radiation reaction, but also the vacuum or free field, which is the solution of the homogeneous quantum-mechanical Maxwell equations. It might be argued that this vacuum field is only of formal significance. In the problem of spontaneous emission, for example, it is necessary to include the vacuum field in the equations of motion for the atomic operators, for otherwise these *operators* would decay to zero as the atom radiates, and we would have a violation of unitarity. The formalism of quantum mechanics demands that we include the vacuum field in the Heisenberg equations of motion.

On the other hand, in the calculation of the Lamb shift the vacuum field seems to be of direct physical significance, at least from one point of view, and indeed Welton showed that it may be considered responsible for the very existence of the Lamb shift. The question then arises whether the vacuum field is physically real or merely an artifice of the formalism.

Consider again the role of the vacuum field in the nonrelativistic theory of the Lamb shift. Welton's argument lends strong support to the idea that the vacuum field is "real," i.e., gives rise to observable effects.

And Power showed explicitly that the Lamb shift may be considered a result of the change in the zero-point energy due to the presence of the atom.[20] (Power's analysis may also be interpreted as showing that the Lamb shift is just the quadratic Stark shift induced by the vacuum field.) Jaynes has considered these calculations of Welton and Power in light of the work of Senitzky and Milonni *et al.*[16]: "This complete interchange-ability of source-field effects and vacuum-fluctuation effects does not show that vacuum fluctuations are 'real.' It shows that source-field effects are the same *as if* vacuum fluctuations were present."[21] Jaynes gives a simple but compelling argument that the effect of radiation re-action is the same as if vacuum fluctuations were present in the problem of spontaneous emission: He shows that the energy density of the radia-tion reaction field is just equal to the energy density of the zero-point field over the spectral range equal to the natural linewidth. "The radiating atom is indeed interacting with an EM field of the intensity predicted by the zero-point energy, but this is just the atom's own radiation reaction field."[21] From this point of view, the interchangeability of the effects of radiation reaction and vacuum fluctuations in spontaneous emission is an example of a fluctuation–dissipation theorem.[21] It seems to the present author that the generalization of these ideas suggested by Jaynes[21] may lead us to view the vacuum field more as a formal artifice or subterfuge than a "real" physical thing.

There has been some very interesting work in the past fifteen years on a purely classical theory of the vacuum field fluctuations. In this work the homogeneous solution of the Maxwell equations is not the null field, which is the conventional *choice*, but rather a field with the energy density

$$(1/8\pi)\langle \mathbf{E}_0^2 + \mathbf{B}_0^2 \rangle = (1/V) \sum_{\mathbf{k},\lambda} \tfrac{1}{2}\hbar\omega_k \qquad (52)$$

Boyer[22] has given an especially appealing formulation of this classical theory by requiring the energy density of the zero-point field to be Lorentz invariant. This requirement leads to the result (52), with $\hbar$ en-tering the theory as a multiplicative constant to be chosen by comparison to results of experiment. In this theory of *random electrodynamics*[22] the vacuum electric field may be expanded in plane waves as

$$\mathbf{E}_0(\mathbf{r}, t) = i \sum_{\mathbf{k},\lambda} (\pi\hbar\omega_k/V)^{1/2} \exp[i(\mathbf{k}\cdot\mathbf{r} - \omega_k t + \theta_{\mathbf{k}\lambda})] \, \hat{\mathbf{e}}_{\mathbf{k}\lambda} + \text{c.c.} \qquad (53)$$

The random property of the field is contained in the $\theta_{\mathbf{k}\lambda}$, which are assumed to be uniformly and independently distributed over the interval $[0, 2\pi]$.

In random electrodynamics the fluctuating vacuum field is consid-ered to be a very real physical thing. In Boyer's derivation of the blackbody spectrum, for example, the zero-point field provides a source of kinetic energy for a particle interacting with radiation.[23] And the black-body spectrum derived by Boyer,

$$\rho_T(\omega) = \frac{\omega^2}{\pi^2 c^3} \left(\frac{\hbar\omega}{\exp(\hbar\omega/kT) - 1} + \tfrac{1}{2}\hbar\omega \right) \tag{54}$$

predicts the zero-point density $\tfrac{1}{2}\hbar\omega$ per mode at $T = 0$, whereas Einstein's approach does not.

There have been some very interesting results from random electrodynamics. Boyer, for instance, has shown that the blackbody spectrum[23] and the van der Waals forces[22] are predicted by random electrodynamics. There is also some interesting work towards uncertainty relations for material systems interacting with the zero-point field.[22] Several other workers have published their views on random or "stochastic" electrodynamics.[24]

6. *Probability Amplitudes and Classical Field Strengths*

It has been recognized from the beginning that one of the most fundamental and peculiar concepts of quantum mechanics is the probability amplitude.[25] Feynman has expressed the view that the assertion of indeterminism was not the most fundamental innovation of quantum theory, that "far more fundamental was the discovery that in nature the laws of combining probabilities were *not* those of the classical probability theory of Laplace." [26] Dirac has taken a similar view[27]:

> The question arises whether the noncommutation is really the main new idea of quantum mechanics. Previously I always thought it was but recently I have begun to doubt it and to think that maybe from the physical point of view, the noncommutation is not the only important idea and there is perhaps some deeper idea, some deeper change in our ordinary concepts which is brought about by quantum mechanics . . . . I believe [the] concept of the probability amplitude is perhaps the most fundamental concept of quantum theory. . . .
>
> The immediate effect of the existence of these probability amplitudes is to give rise to interference phenomena. If some process can take place in various ways, by various channels, as people say, what we must do is to calculate the probability amplitude for each of these channels. Then add all the probability amplitudes, and only after we have done this addition do we form the square of the modulus and get the total result for the probability taking place. You see that that result is quite different from what we should have if we had taken the square of the modulus of the individual terms referring to the various channels. It is this difference which gives rise to the phenomenon of interference, which is all pervading in the atomic world.

I shall now discuss a few examples of the interpretation of radiative processes in terms of probability amplitudes. Once again attention will be focused on the phenomenon of spontaneous emission.

In quantum electrodynamics the radiation from an atom is not a classical field, but instead may be associated with a probability amplitude. The two entities, the classical electromagnetic field amplitude and the quantum-mechanical probability amplitude, both follow the principle of linear superposition, and there is in many cases a strong formal correspondence between them. The best known example of this correspondence is perhaps provided by the Young two-slit experiment.[26] In both the classical and quantum theories a quantity is calculated and then squared to obtain the observed interference pattern. Classically, we may regard each slit as a "source" of radiation, and at any point on the detecting screen the total electric field is just the sum of the fields from these sources; the square of this total field gives the observed intensity pattern. In the quantum-mechanical description the total probability amplitude for photon detection at a point on the detecting screen is the sum of two amplitudes, corresponding to the two (indistinguishable) routes by which the light can reach that point. The probability of photon detection is the square (of the modulus) of the total probability amplitude. The probability distribution on the detecting screen is obtained by repeating the one-photon experiment many times and recording the individual spots on the screen. (An experiment with a many-photon field may be regarded for this purpose as an ensemble of one-photon experiments, since "each photon interferes only with itself. Interference between different photons never occurs."[28]) The correspondence between the classical and quantum descriptions of the single-photon Young experiment has been discussed by Knight and Allen.[29]

The two theories give different predictions for the two-slit experiment when the light is of such low intensity that, quantum mechanically, it can excite only one detector atom (or eject only one photoelectron). Then we record only one spot on the detecting screen, and the calculated interference pattern gives us the probability for the photon to be detected at any given point on the screen. Classically, however, we should always observe the entire interference pattern. In other words, the classical electric field and the quantum-mechanical probability amplitude (or vector potential) have precisely the same spatial distribution on the detecting screen, but the quantum-theoretical probability amplitude gives us the *probability* of detecting one photon at one place. One photon may be associated with a field (spatial) *distribution*. This, of course, is how quantum theory reconciles the wave and particle points of view.

One basic conflict here between the classical and quantum radiation theories is that in the classical theory the field energy that appears in the energy conservation law is

$$H = (1/8\pi) \int d^3r \, [\mathbf{E}(\mathbf{r})^2 + \mathbf{H}(\mathbf{r})^2] \qquad (55)$$

whereas the energetic property of the field in regard to producing excitations in appropriate detector atoms in a semiclassical approach is, as
in quantum electrodynamics, its frequency. Thus, the spatially extended
radiation from a single atom can, according to semiclassical theory, excite
more than one detector atom. In the quantum theory of radiation the
field energy (55) of a monochromatic wave is proportional to the frequency, and a single radiative transition from one atom can excite at
most one other atom.

   The question thus arises whether in nature the radiation from a single
atomic transition can excite more than one other atom. Previous experiments to test this prediction, and a more definitive, sophisticated experiment, have been discussed by Clauser.[30] The semiclassical prediction of
multiatom excitations produced by single-atom radiative decay is in conflict with Clauser's data.

   In our second example, we consider the two three-state schemes
illustrated in Figures 2a and 2b.[31] In case A, represented in Figure 2a,
the atom is supposed to be excited to a coherent superposition of states
2 and 3, both of which are connected to state 1 by allowed transitions.
In case B (Figure 2b), the atom is excited to state 3, which is connected
to the lower states 1 and 2. A semiclassical theory predicts beats in the
radiative emission for both cases A and B. These beats result from the
interference of the fields from the two transition dipoles. Quantum electrodynamics predicts beats in the emission from atoms in case A, the
interference term arising from the addition of probability amplitudes for
the indistinguishable processes $3 \rightarrow 1$ and $2 \rightarrow 1$. In case B, however,
the two probability amplitudes do not interfere, since the (orthogonal)
final states are distinguishable, and no beats are predicted.[31] Semiclassical radiation theory is unable to treat the fundamental distinguishability
of the two final states in case B, and so predicts a result that would be
obtained in the quantum theory if the probability amplitudes were (in-
correctly) added for processes with distinguishable final states.[11]

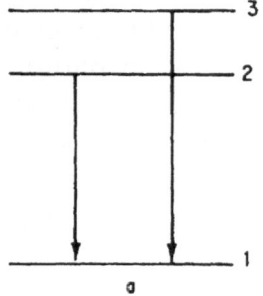 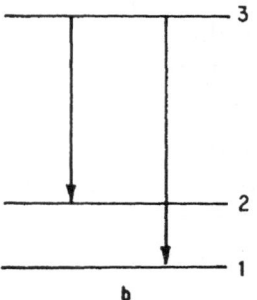

FIGURE 2

Finally, I should like to mention an example in which a classical treatment of the field fails to treat properly sequential processes, where quantum mechanically the probability amplitudes for two successive processes are multiplied to obtain the amplitude for the overall process. This example, the Kocher–Commins experiment,[32] has been discussed from this point of view elsewhere,[11] after Clauser[33] showed that the experimental results could not be explained by conventional semiclassical theories. Therefore, I shall consider only what I perceive to be the essential feature of the experiment that cannot be explained with ordinary classical radiation theory.

Suppose we have a source that produces two photons propagating in opposite directions with the same linear polarization. We do not know the direction of this linear polarization, only that it is the same for the two photons. Each photon must pass through a linear polarizer before photodetection. The two polarizers are oriented at a relative angle ϕ. What is the probability, as a function of ϕ, that a photon passes through each polarizer? In particular, suppose that $\phi = \pi/2$. What is the probability of joint detection in this case? It is easily seen that in this case the probability of joint detection is zero. For if one detector registers a count, the detected photon has linear polarization exactly along the associated polarizer direction. The other photon has this same linear polarization, and so cannot pass through the polarizer it encounters. This is the familiar "all or nothing" nature of photon polarization.

Classically, however, a count at one photodetector does not necessarily mean that the "photon" had polarization such that all of it passed through the polarizer. We can say only that some fraction of the incident radiation intensity passed through the polarizer, in accordance with Malus' law. Thus, a nonvanishing joint probability of photodetection is predicted when $\phi = \pi/2$.

The results of the Kocher–Commins experiment on photon polarization correlations in a three-level cascade cannot be explained by ordinary semiclassical radiation theories based on the Maxwell equations.

7. Concluding Remarks

Spontaneous emission, one of the simplest problems in quantum electrodynamics, and historically its earliest triumph, provides an important testing ground for any theory of light and matter. In it we find some of the most profound aspects of contemporary radiation theory, such as the vacuum field fluctuations, radiation reaction, renormalization, and photon interference effects.

There is no doubt that the quantum theory of radiation contains many important elements of truth. The papers at this symposium prove, however, that not everyone is entirely complacent. We can thank people like Professor Jaynes for helping to maintain this healthy situation.

References and Notes

1. P. A. M. Dirac, *Proc. R. Soc. London Ser. A* **114**, 243 (1927).
2. W. Heisenberg, *Z. Phys.* **33**, 879 (1925).
3. R. Ladenburg, *Z. Phys.* **4**, 451 (1921).
4. A. Einstein, *Phys. Z.* **18**, 121 (1917).
5. W. Kuhn, *Z. Phys.* **33**, 408 (1925).
6. M. Born and P. Jordan, *Z. Phys.* **34**, 858 (1925).
7. H. A. Kramers, *Nature (London)* **133**, 673 (1924).
8. R. Ladenburg and H. Kopfermann, *Nature (Paris)* **122**, 438 (1928).
9. M. Born, W. Heisenberg, and P. Jordan, *Z. Phys.* **35**, 557 (1926).
10. S. Weinberg, *Proc. Am. Acad. Arts Sci.* **106**, 22 (1977).
11. See P. W. Milonni, *Phys. Lett. C* **25**, 1 (1976), and references therein to work of J. R. Ackerhalt, P. L. Knight, and J. H. Eberly.
12. H. A. Bethe, *Phys. Rev.* **72**, 339 (1947).
13. W. E. Lamb, Jr. and R. C. Retherford, *Phys. Rev.* **72**, 241 (1947).
14. See, for instance, footnote 13 of R. P. Feynman, *Phys. Rev.* **76**, 769 (1949).
15. T. A. Welton, *Phys. Rev.* **74**, 1157 (1948).
16. This view of the roles of radiation reaction and vacuum field fluctuations in spontaneous emission was arrived at independently and simultaneously by I. R. Senitzky and P. W. Milonni, J. R. Ackerhalt, and W. A. Smith (see reference 11).
17. C. R. Stroud, Jr. and E. T. Jaynes, *Phys. Rev. A* **1**, 106 (1970).
18. J. A. Wheeler and R. P. Feynman, *Rev. Mod. Phys.* **17**, 157 (1945).
19. See, however, F. Hoyle and J. V. Narlikar, *Action at a Distance in Physics and Cosmology*, Freeman, San Francisco (1974).
20. E. A. Power, *Am. J. Phys.* **34**, 516 (1966).
21. E. T. Jaynes, *Coherence and Quantum Optics*, Eds. L. Mandel and E. Wolf, Plenum, New York (1978). pp. 495–509.
22. See, for example, T. H. Boyer, *Phys. Rev. D* **11**, 790 (1975).
23. T. H. Boyer, *Phys. Rev.* **182**, 1374 (1969).
24. See, for example, the following papers and references therein: E. Santos, *Nuovo Cimento B* **22B**, 201 (1974); P. Claverie and S. Diner, in *Localization and Delocalization in Quantum Chemistry*, Vol II, Eds. O. Chalvet *et al.*, Reidel Publishing Co., Dordrecht, Holland (1976), pp. 395–448; L. de la Peña-Auerbach and A. M. Cetto, *Found. Phys.* **8**, 191 (1978).
25. See, for example, W. Heisenberg, *The Physical Principles of the Quantum Theory*, Dover, New York (1949).
26. R. P. Feynman, *Proceedings of the Second Berkeley Symposium on Mathematical Statistics and Probability*, University of California Press, Berkeley (1951), p. 533.
27. P. A. M. Dirac, *Fields Quanta* **3**, 154 (1972).
28. P. A. M. Dirac, *Principles of Quantum Mechanics*, 4th ed., Oxford University Press, London (1958), p. 9.
29. P. L. Knight and L. Allen, *Opt. Commun.* **7**, 44 (1973).
30. J. F. Clauser, *Phys. Rev. D* **9**, 853 (1974).

31. W. W. Chow, M. O. Scully, and J. O. Stoner, Jr., *Phys. Rev. A* **11,** 1380 (1975); R. M. Herman, H. Grotch, R. Kornblith, and J. H. Eberly, *Phys. Rev. A* **11,** 1389 (1975).
32. C. A. Kocher and E. D. Commins, *Phys. Rev. Lett.* **18,** 575 (1967).
33. J. F. Clauser, *Phys. Rev. A* **6,** 49 (1972).

2

Unified View of Spontaneous Emission in Several Theories of Radiation

J. H. Eberly

1. Introduction

Consider the model classical field theory described by the Lagrangian density $\mathscr{L}(\mathbf{x},t)$:

$$\mathscr{L}(\mathbf{x},t) = \psi^*(\mathbf{x},t)\left\{ i\hbar\frac{\partial}{\partial t} - \frac{1}{2m}\left[\frac{\hbar}{i}\boldsymbol{\nabla} - \frac{e}{c}\mathbf{A}(\mathbf{x},t)\right]^2 - V(\mathbf{x})\right\}\psi(\mathbf{x},t)$$

$$+ \frac{1}{8\pi}\left\{\left[\frac{1}{c}\dot{\mathbf{A}}(\mathbf{x},t)\right]^2 - [\boldsymbol{\nabla}\times\mathbf{A}(\mathbf{x},t)]^2\right\} \tag{1}$$

Evidently, $\mathscr{L}(\mathbf{x})$ is the beginning of a theory of the interaction of the fields ψ and $\mathbf{A}$. There are several free parameters in the theory: $\hbar$, m, e, c.

The functional derivative of $\mathscr{L}$ with respect to $\dot{\psi}$ and $\dot{\mathbf{A}}$ shows that the conjugate momenta Π and $\mathscr{P}$ are

$$\Pi(\mathbf{x},t) = i\hbar\psi^*(\mathbf{x},t) \tag{2}$$

$$\mathscr{P}(\mathbf{x},t) = (4\pi c^2)^{-1}\dot{\mathbf{A}}(\mathbf{x},t) \tag{3}$$

The Hamiltonian therefore has the form

$$H = \int d^3x\left\{\psi^*\left[\frac{1}{2m}\left(\frac{\hbar}{i}\boldsymbol{\nabla} - \frac{e}{c}\mathbf{A}\right)^2 + V(\mathbf{x})\right]\psi\right.$$

$$\left. + \frac{1}{8\pi}\left[(4\pi c\mathscr{P})^2 + (\boldsymbol{\nabla}\times\mathbf{A})^2\right]\right\} \tag{4}$$

where ψ^* has been used in place of $(i\hbar)^{-1}\Pi$ for simplicity.

J. H. Eberly • Department of Physics and Astronomy, University of Rochester, Rochester, New York 14627. This work was performed while the author was a Visiting Fellow at the Joint Institute for Laboratory Astrophysics, University of Colorado and National Bureau of Standards, Boulder, Colorado 80309.

It is clear that **A** is intended to be the electromagnetic vector potential field, and that ψ is some kind of matter field. It is important to recognize that, despite the appearance of $\hbar$, our model theory is so far completely classical.[1] The equations of motion are simply Hamilton's equations. If $\mathscr{F}$ is any dynamical field, then

$$\dot{\mathscr{F}} = [\mathscr{F}, H]_{\text{PB}} \tag{5}$$

where PB means classical Poisson bracket. In the present case this means

$$\dot{\mathscr{F}}(\mathbf{x},t) = \int d^3 y \left(\frac{\delta \mathscr{F}}{\delta \psi(\mathbf{y},t)} \frac{\delta H}{\delta \Pi(\mathbf{y},t)} - \frac{\delta \mathscr{F}}{\delta \Pi(\mathbf{y},t)} \frac{\delta H}{\delta \psi(\mathbf{y},t)} \right)$$
$$+ \int d^3 y \left(\frac{\delta \mathscr{F}}{\delta \mathbf{A}(y,t)} \frac{\delta H}{\delta \mathscr{P}(y,t)} - \frac{\delta \mathscr{F}}{\delta \mathscr{P}(y,t)} \frac{\delta H}{\delta \mathbf{A}(y,t)} \right) \tag{6}$$

and $\delta/\delta\psi$, $\delta/\delta\mathbf{A}$, etc., denote functional derivatives of the usual kind.

If we substitute ψ itself for $\mathscr{F}$ we find a familiar equation:

$$i\hbar\dot{\psi} = \left[\frac{1}{2m} \left(\frac{\hbar}{i} \mathbf{\nabla} - \frac{e}{c} \mathbf{A} \right)^2 + V(\mathbf{x}) \right] \psi \tag{7}$$

This is Schrödinger's equation if ψ has its usual wave-mechanical interpretation. Let us not assume any such interpretation but simply continue to follow the dictates of the original Lagrangian. We will, however, use mathematical procedures that are familiar from wave mechanics. Let us expand ψ in the basis of the eigenstates of equation (7) with $\mathbf{A} = 0$, that is, in the decoupled eigenbasis:

$$\psi(\mathbf{x},t) = \sum_{\mathbf{p}} b_{\mathbf{p}}(t) \mathscr{U}_{\mathbf{p}}(\mathbf{x}) \tag{8}$$

If the $\mathscr{U}$'s are orthonormal it follows that

$$b_{\mathbf{p}}(t) = \int d^3 x \, \mathscr{U}_{\mathbf{p}}^*(\mathbf{x})\psi(\mathbf{x},t) \tag{9}$$

and that

$$[b_{\mathbf{p}}(t), i\hbar b_{\mathbf{q}}^*(t)]_{\text{PB}} = \delta(\mathbf{p},\mathbf{q}) \tag{10}$$

In other words, transformation (8) is canonical, and $b_{\mathbf{p}}$ and $i\hbar b_{\mathbf{p}}^*$ are the new field and conjugate momentum variables. The index $\mathbf{p}$ may be both discrete and continuous in its range.

At the same time, the decoupled eigensolutions of **A** and $\mathscr{P}$ can be taken to be $V^{-1/2} \exp[\pm i\mathbf{k}\cdot\mathbf{x}]$, as is well known, so we write

$$\mathbf{A}(\mathbf{x}, t) = \sum_{\lambda} \left(\frac{\hbar c^2}{2\omega_\lambda V} \right)^{1/2} \boldsymbol{\epsilon}_\lambda \left[a_\lambda(t)\exp(i\mathbf{k}_\lambda\cdot\mathbf{x}) + a_\lambda^*(t)\exp(-i\mathbf{k}_\lambda\cdot\mathbf{x}) \right] \tag{11}$$

$$\mathscr{P}(\mathbf{x},\,t) = -i \sum_{\lambda} \left(\frac{\hbar\omega_{\lambda}}{2c^2V} \right)^{1/2} \epsilon_{\lambda}[a_{\lambda}(t)\exp(i\mathbf{k}_{\lambda}\cdot\mathbf{x}) - a^*_{\lambda}(t)\exp(-i\mathbf{k}_{\lambda}\cdot\mathbf{x})] \quad (12)$$

There is little motivation here for the square roots, but we anticipate they will have ultimate utility if this theory is to have any connection with atomic radiation physics. The $\hbar$'s are placed conventionally, but we notice additionally that they imply the classical relation

$$[a_{\lambda},\, i\hbar a^*_{\lambda'}]_{\text{PB}} = \delta(\lambda,\,\lambda') \quad (13)$$

which is the analog of (10).

Now we write the entire Hamiltonian in terms of the new canonical fields. We find

$$H = \sum_{\mathbf{p}} E_{\mathbf{p}} b^*_{\mathbf{p}} b_{\mathbf{p}} + \sum_{\lambda} \hbar\omega_{\lambda} a^*_{\lambda} a_{\lambda} + i\hbar \sum_{\mathbf{pq}\lambda} b^*_{\mathbf{p}} b_{\mathbf{q}} (M_{\mathbf{pq}\lambda} a_{\lambda} - M^*_{\mathbf{qp}\lambda} a^*_{\lambda}) \quad (14)$$

where the parameter M is given by

$$M_{\mathbf{pq}\lambda} = \frac{e}{mc} \left(\frac{\hbar c^2}{2\omega_{\lambda} V} \right)^{1/2} \int d^3x\, \mathscr{U}^*_{\mathbf{p}}(\mathbf{x})\exp(i\mathbf{k}_{\lambda}\cdot\mathbf{x})\epsilon_{\lambda}\cdot\boldsymbol{\nabla}\, \mathscr{U}_{\mathbf{q}}(\mathbf{x}) \quad (15)$$

This form of M ignores the $\mathbf{A}^2$ term in the Hamiltonian altogether for the sake of brevity.

Now, because $b_{\mathbf{p}}$ and a_{λ} are canonical fields and $i\hbar b^*_{\mathbf{p}}$ and $i\hbar a^*_{\lambda}$ are their conjugate momenta, we can obtain a complete set of equations for the coupled classical fields from the Poisson brackets, using the H in (14). For example,

$$i\hbar\dot{a}_{\lambda} = [a_{\lambda},\, H]'_{\text{PB}}$$

$$= \sum_{\mu} \left(\frac{\delta a_{\lambda}}{\delta a_{\mu}} \frac{\delta H}{\delta a^*_{\mu}} - \frac{\delta a_{\lambda}}{\delta a^*_{\mu}} \frac{\delta H}{\delta a_{\mu}} \right) + \sum_{\mathbf{q}} \left(\frac{\delta a_{\lambda}}{\delta b_{\mathbf{q}}} \frac{\delta H}{\delta b^*_{\mathbf{q}}} - \cdots \right)$$

$$= \hbar\omega_{\lambda} a_{\lambda} - i\hbar \sum_{\mathbf{pq}} M^*_{\mathbf{qp}\lambda} b^*_{\mathbf{p}} b_{\mathbf{q}} \quad (16)$$

Here the prime on the bracket merely means that it has been multiplied by $i\hbar$ or, in other words, that the bracket is with respect to a^*_{λ}, $b^*_{\mathbf{p}}$ instead of the true momenta, $i\hbar a^*_{\lambda}$, $i\hbar b^*_{\mathbf{p}}$. Similarly, one finds

$$i\hbar\dot{b}_{\mathbf{p}} = [b_{\mathbf{p}},\, H]'_{\text{PB}} = E_{\mathbf{p}} b_{\mathbf{p}} + i\hbar \sum_{\mathbf{rs}\lambda} b_s \delta_{\text{pr}}(M_{\mathbf{rs}\lambda} a_{\lambda} - M^*_{\mathbf{sr}\lambda} a^*_{\lambda}) \quad (17)$$

2. Quantization and QED

Having come so far it is appropriate to introduce quantization in the ordinary way. We simply assert that the canonical variables are operators in a vector space, and that they do not commute, but obey certain definite commutation relations. In the case of the a's, we impose the relations

$$[a_\lambda, a_{\lambda'}^\dagger] = \delta_{\lambda\lambda'} \qquad [a_\lambda, a_{\lambda'}] = [a_\lambda^\dagger, a_{\lambda'}^\dagger] = 0 \qquad (18)$$

which are certainly familiar. In the case of the b's we must impose *anti*commutation rules in order to take account of the Pauli exclusion principle that affects electrons and all other fermions. For the b operators we have the rules

$$\{b_p, b_q^\dagger\} = \delta_{pq} \qquad \{b_p, b_q\} = \{b_p^\dagger, b_q^\dagger\} = 0 \qquad (19)$$

The eigenstates of the unperturbed electron and electromagnetic field Hamiltonians are described as usual. A general electron state is a sum of no-electron states, one-electron states, two-electron states, and so on. In the same way, the general photon state is a sum of zero-, one-, and all higher-number photon states. For example, such a photon state might be written

$$| \Phi\{n_\lambda\}\rangle = \Bigg[c_0 + \sum_{\lambda_1} c(\lambda_1) a_{\lambda_1}^\dagger $$
$$+ \tfrac{1}{2} \sum_{\lambda_1} \sum_{\lambda_2} c(\lambda_1\lambda_2) a_{\lambda_1}^\dagger a_{\lambda_2}^\dagger + \cdots \Bigg] |\{0\}\rangle \qquad (20)$$

where $|\{0\}\rangle$ is the "vacuum" state, the state for which $a_\lambda|\{0\}\rangle = 0$, for all modes λ. In the case of electrons, our interest in atomic physics allows some simplification, because we can restrict our attention to single-electron atoms and still have a great deal of physics to understand. We will make that simplification in the following paragraphs, understanding that we have truncated the fermion half of our theory, not allowing ourselves the luxury of multielectron atoms.

Of course, the theory of atomic electrons can be formulated in a much more complete way by adopting a slightly deeper approach in which a closed atomic shell is viewed as a kind of "vacuum," and the valence electrons comprise the positive-energy excitations. The possibility of exciting additional electrons out of the closed shell is accommodated by including fermion "hole" operators into the theory, and so on.

When we wrote the general photon state (20) we left time out of the coefficients, preferring to think of the operators as changing with time. This is the simplest classical point of view; the dynamical variables change with time. The variables are now operators, and the equations of motion that they obey must take careful account of possible noncommutativity. Nevertheless, the equations for any operator G are easily derived using the Heisenberg prescription:

$$i\hbar\dot{G} = [G, H] \qquad (21)$$

In the long-wavelength limit, when the factor $\exp(i\mathbf{k}_\lambda \cdot \mathbf{x})$ can be replaced by unity in the $M_{\mathbf{pq}\lambda}$ integral (15), the equation for the field mode operator a_λ is found to be

$$\dot{a}_\lambda = -i\omega_\lambda a_\lambda + \sum_{\mathbf{pq}} M_{\mathbf{pq}\lambda} b_\mathbf{p}^\dagger b_\mathbf{q} \tag{22}$$

In order to find the electron operator equations it is useful to note that only pairs of b's and $b^\dagger$'s enter the Hamiltonian. One usually says that the pair $b_\mathbf{r}^\dagger b_\mathbf{s}$ destroys an electron in state $\mathbf{s}$ and creates one in state $\mathbf{r}$. In our one-electron atom this is the same as making a transition from $\mathbf{s}$ to $\mathbf{r}$, so we will denote the product $b_\mathbf{r}^\dagger b_\mathbf{s}$ by the transition operator $T_{\mathbf{rs}}$:

$$T_{\mathbf{rs}} = b_\mathbf{r}^\dagger b_\mathbf{s} \quad [\text{QED}] \tag{23a}$$

and then look for the equations which the transition operators obey. From relations (19) we find

$$[T_{\mathbf{pq}}, T_{\mathbf{rs}}] = T_{\mathbf{ps}}\delta_{\mathbf{rq}} - T_{\mathbf{rq}}\delta_{\mathbf{ps}} \tag{23b}$$

It then follows directly from (21) that we have:

$$\dot{T}_{\mathbf{rs}} = i\omega_{\mathbf{rs}} T_{\mathbf{rs}} + \sum_{\mathbf{pq}\lambda} M_{\mathbf{pq}\lambda}(T_{\mathbf{rq}}\delta_{\mathbf{ps}} - T_{\mathbf{ps}}\delta_{\mathbf{rq}})(a_\lambda + a_\lambda^\dagger) \tag{24}$$

where we have used the standard transition frequency notation in which $\omega_{\mathbf{rs}}$ means $(E_\mathbf{r} - E_\mathbf{s})/\hbar$.

3. QED Source Fields: Decay Rates and Level Shifts

We want to exhibit some nontrivial solutions to these very complicated operator equations (22) and (24), so some simplification is necessary. We can achieve adequate simplicity by adopting a specific context. We have chosen the context of atomic emission and absorption of light. To an excellent approximation only a single transition needs to be considered in order to discuss the main elements of emission and absorption. Thus we simply restrict the indices $\mathbf{r}$, $\mathbf{s}$, $\mathbf{p}$, $\mathbf{q}$ to the values 1 and 2. This is reminiscent of, but not equivalent to, the "essential states" restriction of Weisskopf and Wigner.[2]

As is well known, in the two-level space the Pauli matrices offer a familiar notation that covers all possibilities. In Pauli matrix language the operator equations of motion are

$$\dot{a}_\lambda = -i\omega_\lambda a_\lambda - M_\lambda(\sigma^\dagger - \sigma) \tag{25a}$$

$$\dot{\sigma} = -i\omega_0\sigma + \sum_\lambda M_\lambda \sigma_3(a_\lambda + a_\lambda^\dagger) \tag{25b}$$

$$\dot{\sigma}_3 = -2\sum_\lambda M_\lambda(\sigma + \sigma^\dagger)(a_\lambda + a_\lambda^\dagger) \tag{25c}$$

where ω_0 is ω_{21}, the (positive) transition frequency, and $M_\lambda = M_{12\lambda} = -M_{21\lambda}$, which is real on the assumption that we have a $\Delta m = 0$ transition to work with. The new operators are related to the transition operators T_{rs} as follows:

$$\sigma = T_{12} \qquad \sigma^\dagger = T_{21} \qquad \sigma_3 = T_{22} - T_{11} \qquad (26)$$

The appearance of σ_3 on the right side of (25b) is our assurance that the two-level restriction is not a trivial linearization. We have arrived at equations slightly more general than those used by Weisskopf and Wigner[2] to discuss spontaneous decay and natural line shape. This is because equations (25) contain the non-energy-conserving virtual transitions discarded in reference 2.

To the extent that the σ's operate on the same state vectors as the $b^\dagger$ and b operators, we can say that even the relatively simple equations (25) represent the dynamics of a *quantum field theory*. However, it is not possible to reconstruct the fermion field operator $\psi(\mathbf{x}, t)$ itself from the solutions to those equations. For that one needs the solutions for the basic operators $b_r^\dagger$ and b_s, not just for their bilinear products. The construction of $\psi(\mathbf{x}, t)$ can be regarded as an exercise to be undertaken for fun but not much profit. As we have seen, all of the operators that appear to have any physics in them, e.g., all of the terms in the Hamiltonian, are bilinear operators.

Ackerhalt and I have explored in some detail the way in which the formalism outlined here can be applied to a more realistic atom model.[3] In particular, we have studied, from the point of view of the dynamics of the bilinear transition operators, the complex radiation frequency shift of the first transition from the ground state of a nonrelativistic hydrogen atom. A similar formalism can be developed[4] to treat free-electron (rather than atomic-electron) problems such as the low-energy theorem, the electron self-mass and runaway, and the anomalous magnetic moment.

The first step is to "solve" the field equation by rewriting it as an integral:

$$a_\lambda = a_\lambda^h - M_\lambda \int_0^t dt' \, [\sigma^\dagger(t') - \sigma(t')] \exp[-i(\omega_\lambda - i\epsilon)(t - t')] \quad (27)$$

where $a^h \equiv a_\lambda(0)\exp(-i\omega_\lambda t)$ is the homogeneous solution, and the ϵ in the exponent reminds us that we have chosen the retarded solution of the Maxwell wave equation. Next, this solution and its adjoint are substituted into the normally ordered versions of (25b) and (25c). This gives equations for the transition operators alone, with the known homogeneous solution of (27) providing time-dependent external forces. For example, the equation for $T_{12} \equiv \sigma$ is

$$\dot{\sigma} = -i\omega_0\sigma + \sum_\lambda M_\lambda[\sigma_3 a_\lambda^h + a_\lambda^{\dagger h}\sigma_3]$$

$$+ \sum_\lambda M_\lambda^2 \int_0^t dt'\sigma_3(t)\sigma(t')\exp[-i(\omega_\lambda - i\epsilon)(t - t')]$$

$$- \sum_\lambda M_\lambda^2 \int_0^t dt'\sigma(t')\sigma_3(t)\exp[i(\omega_\lambda + i\epsilon)(t - t')] \qquad (28)$$

So-called counter-rotating terms have been kept in the solution of (27), but discarded after the substitution into (25b). The implications of this step have been discussed.[3]

A few remarks about (28) are perhaps useful. All of the terms contain operators. These operators obey the equal-time commutation relations given in (18) and either (19) or (23). Equation (28) is fully quantum electrodynamic in the context of the present nonrelativistic and dipole-approximate model. It is clear why the procedure outlined here is called QED source-field theory, or operator radiation reaction theory. Several overviews of recent work based on this procedure are available[5] and these can be consulted for references to the original papers.

To finish our QED discussion we must return to (28) and solve it. By this we mean that we adopt the usual quantum meanings of the terms and symbols and solve for the appropriate expectation values, because it is these that are assumed by QED to be measured in experiments. The first terms of nontrivial content are of order M^2. We will not keep higher orders in the equation. A simple approach for removing higher-order terms, but one burdened by Markovian assumptions, is explained in the next paragraph. A better but much more elaborate approach is easy to explain. As Wódkiewicz has pointed out, because σ_3 is constant to order M, we change nothing through order M^2 by replacing $\sigma_3(t)$ by $\sigma_3(t')$ in (28). Then the expectation (in the vacuum field state) of (28) becomes

$$\langle\dot{\sigma}\rangle = -i\omega_0\langle\sigma\rangle - \sum_\lambda M_\lambda^2 \int_0^t dt' \langle\sigma(t')\rangle$$

$$\times \{\exp[-i\omega_\lambda(t - t')] + \exp[i\omega_\lambda(t - t')]\}$$

where we have used $\sigma_3\sigma = -\sigma$ and similar relations that follow from (26), as well as the vacuum identities $\langle\sigma_3 a_\lambda^h\rangle = 0$ and $\langle a_\lambda^{\dagger h}\sigma_3\rangle = 0$. The important point is that the equation above is a *linear* Volterra integral equation that can be solved straightforwardly and exactly by Laplace transform methods. The correlation function $\langle\sigma^\dagger(t + \tau)\sigma(t)\rangle$, which is more interesting from the point of view of the radiated spectrum, can be found in the same way. The spectrum is not Lorentzian, and the decay law for the atom is nonexponential. The deviations are very small and are already known in the context of a careful solution to the Weisskopf–Wigner problem.[2,6]

In (28) the operator products $\sigma_3(t)\sigma^\dagger(t')$, etc., can also be treated with a simple approximation. The lowest-order solutions for the rapidly varying operators $\sigma(t')$ and $\sigma^\dagger(t')$ can be used:

$$\sigma(t') \simeq \sigma(t)\exp[i\omega_0(t - t')] \qquad \sigma^\dagger(t') \equiv \sigma^\dagger(t)\exp[-i\omega_0(t - t')] \quad (29)$$

Then the operator parts do not depend on t' and are taken out of the integral. This leaves behind an integral which can be expressed, after sufficient time, in terms of delta functions and principal parts. In place of (28), one gets the important relation

$$\dot\sigma = -i\omega_0\sigma + \sum_\lambda M_\lambda(\sigma_3 a_\lambda^h + a_\lambda^{\dagger h}\sigma_3)$$

$$+ \sum_\lambda M_\lambda^2[(\pi\delta - iP)^{(-)}\sigma_3\sigma - (\pi\delta + iP)^{(+)}\sigma\sigma_3] \qquad (30)$$

The superscript $(\pm)$ means that the argument of the singular functions is to be $\omega_\lambda + \omega_0$ or $\omega_\lambda - \omega_0$. The use of the identities $\sigma_3\sigma = -\sigma$ and $\sigma\sigma_3 = \sigma$ reduces (30) to

$$\dot\sigma = -i\omega_0\sigma + \sum_\lambda M_\lambda(\sigma_3 a_\lambda^h + a_\lambda^{\dagger h}\sigma_3)$$

$$+ \left[\sum_\lambda M_\lambda^2 \left(-\pi\delta(\omega_\lambda - \omega_0) + \frac{iP}{\omega_\lambda - \omega_0} - \frac{i}{\omega_\lambda + \omega_0} \right) \right] \sigma$$

or

$$\langle\dot\sigma\rangle = -i\omega_0\langle\sigma\rangle - (\tfrac{1}{2}A + i\Delta)\langle\sigma\rangle \qquad (31)$$

Here the decay rate is

$$A = 2\pi \sum_\lambda M_\lambda^2 \delta(\omega_\lambda - \omega_0) \qquad (32a)$$

and the frequency shift is

$$\Delta = -\sum_\lambda M_\lambda^2 \left(\frac{P}{\omega_\lambda - \omega_0} - \frac{1}{\omega_\lambda + \omega_0} \right) \qquad (32b)$$

Several comments are relevant. The decay rate A is exactly the Einstein A coefficient as it should be. The frequency shift Δ is logarithmically divergent as it should be. The normal ordering we have chosen allows the result to be obtained most simply. Other orderings are not quite so straightforwardly handled, but give the same answers. These points have been discussed.[7] What is most relevant here, and should be emphasized, is that the homogeneous parts of the solutions to equations (25), very troublesome terms, do not appear in the final result (31). Note that they have been eliminated not by any mathematical device, but *by interpretation*. That is, in the quantum theory the interpretive framework demands, for spontaneous emission, a vacuum expectation value. It is this demand that eliminates the homogeneous solutions.

4. Random Electrodynamics

The name of this second field theory is offered with apologies to Boyer who has investigated a nonquantized theory of electrodynamics in the past several years,[8] a theory which he refers to as random electrodynamics (RED). Although the field theory which I'll discuss here is not identical with Boyer's, the similarities are very great, and the existence of Boyer's work is my main motivation. Only an inspection of Boyer's own papers gives an adequate picture of the scope of his theory.

My version of RED starts with the Lagrangian and Hamiltonian given in (1) and (4). Because nothing is quantized we can omit all the steps taken at the beginning of Section 2 and work immediately with the classical equations (16) and (17) given at the end of Section 1. The Hamiltonian suggests that bilinear combinations of b^*'s and b's may be as convenient to work with in RED as in QED. From (16) and (17) we easily find that the QED equations (25) *are exactly true classically*. However, the "transition operators" T_{pq} are no longer operators on a vector space but just complex numbers:

$$T_{pq} = b_p^* b_q \qquad \text{[RED]} \qquad (33)$$

Let us look at solutions to equations (25) without imputing any operator character to the variables. We will make the same numerical approximations made for the QED solutions: only terms through order M^2 are to be kept, and the Markovian assumption will again be used. An integral equation for σ is again obtained, this time after the solutions for σ_3, a_λ, and a_λ^* are inserted on the right side of (25b):

$$\dot{\sigma} = -i\omega_0\sigma - \sum_\lambda M_\lambda \left[\sigma_3(0) + 2 \sum_\mu M_\mu \int dt' \, (a_\mu + a_\mu^*)^h (\sigma + \sigma^*) \right]$$

$$\times \left[(a_\lambda + a_\lambda^*)^h + M_\lambda(\text{terms linear in } \sigma^h \text{ and } \sigma^{h\dagger}) \right] \qquad (34)$$

As before, the superscript h means the homogeneous solution for the variable.

At this point we introduce a stochastic postulate into the theory. It is motivated by, but is more demanding than, a similar postulate introduced by Boyer in his work.[8] Let us assert that homogeneous solutions to our electromagnetic and electron field equations are only stochastically knowable. That is, only averages of our solutions can be compared with experiment. Note that this is an *interpretive statement* imposed from the outside and not a mathematical consequence of anything. Furthermore, we assert that the equal-time homogeneous solutions for the σ, σ^*, a_λ, a_λ^* variables are uncorrelated with each other as if they all carried random phases:

$$[\sigma\sigma]^h = [\sigma\sigma_3]^h = [a_\lambda a_\mu]^h = \cdots = 0 \qquad [a_\lambda a_\mu^*]^h = [a_\mu^* a_\lambda]^h = \tfrac{1}{2}\delta_{\lambda\mu} \quad (35)$$

where the square brackets denote an average over the underlying statistical ensemble. For our purposes we do not need to specify $[\sigma\sigma^*]$.

The significance of (34) is hidden in its average value, according to our postulate. The equation simplifies a great deal after the average is taken:

$$[\dot\sigma] = -i\omega_0[\sigma] - [\sigma] \sum_\lambda M_\lambda^2 \int_0^t dt'$$

$$\times \{\exp[-i(\omega_\lambda - \omega_0)(t - t')] + \exp[i(\omega_\lambda + \omega_0)(t - t')]\} \quad (36)$$

We have already encountered these same singular exponential integrals in Section 3. When we evaluate them we find that their effect is to add a complex shift to the frequency ω_0. In fact, this complex shift is *exactly* the same as that found in QED. Its real and imaginary parts are just Δ and A given in equations (32). Clearly RED doesn't differ from QED at all in its second-order predictions of decay rates and level shifts.

5. Neoclassical Electrodynamic Field Theory

Since 1956 Jaynes[9–11] has been considering the possibility of constructing a quantum-mechanical theory of electrodynamics in which the field variables would not be operators, and the atomic variables would not be second quantized but would be interpretable in wave-mechanical terms. In 1972 he summarized[10] the state of development of a theory of that kind called neoclassical theory (NCT). Mandel[12] has recently reviewed the comparative standing of NCT and QED in light of recent experiments. Although Jaynes has made no final claims for it, NCT has become unusually controversial. In the succeeding paragraphs we discuss a similar theory[13] called neoclassical electrodynamics (NED). We will show that NED is just like QED and RED in one very basic respect: it can be regarded as an attempt to interpret the field theory underlying the Lagrangian (1) and the Hamiltonian (4).

NED asserts that all of the fields in the theory are nonoperator fields, and that the equations of motion (25) are nonoperator equations. In this respect NED is exactly like RED. However, NED asserts that homogeneous solutions are knowable functions of space and time. In particular, the homogeneous solutions for the field operators a_λ and a_λ^* need not be treated stochastically and indeed must be taken to be zero in the absence of external sources. This has the effect, when considering spontaneous emission, of simplifying the basic equation (28) to

$$\dot{\sigma} = -i\omega_0\sigma - \sum_\lambda M_\lambda^2 \int_0^t dt' \{-\sigma_3(t)\sigma(t')$$

$$\times \exp[-i\omega_\lambda(t - t')] + \sigma(t')\sigma_3(t) \exp[i\omega_\lambda(t - t')]\} \quad (37)$$

By now we are familiar with the steps to be taken to complete the solution of (37). The homogeneous solution for $\sigma(t')$ is permissible, in light of the M^2 multiplying the integrals: $\sigma(t') \simeq \sigma(t)\exp[i\omega_0(t - t')]$. Thus, (37) simplifies to

$$\dot{\sigma} = -i\omega_0\sigma - \sum_\lambda M_\lambda^2(\cdots) \quad (38a)$$

$$(\cdots) \equiv -\sigma_3\sigma \int dt' \exp[-i(\omega_\lambda - \omega_0)(t - t')] + \sigma\sigma_3$$

$$\times \int dt' \exp[i(\omega_\lambda + \omega_0)(t - t')] \quad (38b)$$

This equation is almost the same as equation (31) with the homogeneous terms thrown away. We could still revert to QED at this point by using the QED relations $\sigma_3\sigma = -\sigma$ and $\sigma\sigma_3 = \sigma$, and the outcome would be exactly (31). However, NED is thoroughly classical as a field theory, so $\sigma_3\sigma = \sigma\sigma_3$, and the NED counterpart of (31) is

$$\dot{\sigma} = -i\omega_0\sigma + (\tfrac{1}{2}A + i\Delta_{\text{NCT}})\sigma_3\sigma \quad (39)$$

where the coefficients are

$$A = 2\pi \sum_\lambda M_\lambda^2\delta(\omega_\lambda - \omega_0) \quad (40a)$$

$$\Delta_{\text{NED}} = -\sum_\lambda M_\lambda^2\left(\frac{P}{\omega_\lambda - \omega_0} + \frac{1}{\omega_\lambda + \omega_0}\right) \quad (40b)$$

Here in NED, as in NCT, $\sigma_3(t)$ modulates the complex frequency shift. When the atom is near its ground state the usual NCT interpretation of σ_3 gives $\sigma_3 \simeq -1$, and the decay of the oscillation of σ occurs at exactly the QED rate. (When the atom is more nearly in its upper state, then σ_3 is positive.) This allows NED and NCT to agree with the results of practically all prelaser spectroscopy, since conventional light sources are not intense enough to excite an ensemble of atoms very far from their ground states.

The matter of the frequency shift Δ_{NED} is more interesting, because it is very different from Δ, the QED shift. The expressions (32b) and (40b) differ in the sign of the second term—the term that can be recognized as coming from "counter-rotating" terms in the Hamiltonian. (In the full rotating-wave approximation, NED and QED agree as to the magnitude of the frequency shift. However, this agreement must not be emphasized because the counter-rotating terms are critically important in any frequency shift calculation.[14] It is apparent at a glance that Δ and Δ_{NED} are possibly dramatically different in their numerical values. While both expressions exhibit ultraviolet divergences and require cutoffs, the degree of divergence of Δ is only logarithmic and that of Δ_{NED} is linear.

6. Remarks

Although none of the results stated here are new, I thought it might be worthwhile to show them again, all derived from a common Hamiltonian. It should be clear, first of all, that (1) implies a rich variety of field theories. Second, by deriving everything from the same starting point, and by making the same numerical approximations every time, we find an explicit solution and we obtain some insights into some of the claims and counterclaims made by advocates of QED, RED, and NED. It should be apparent that both RED and NED are legitimate field theories, despite the differences between them, and their differences from canonical QED. At the level presented here they are obviously internally consistent and susceptible to expansion outside the dipole approximation, and outside the nonrelativistic limit, in ways that parallel the familiar path that takes QED into the relativistic domain.

In the end there is the question of correctness. Of course, it is the fate of *every* theory to be wrong in the long run. But, for the moment, what theory of electrodynamics is closest to being right? That is a much more slippery question than generally acknowledged. In order to answer it fairly one must remember that a theory can be compared with experiments only by using the theory's own interpretive framework. Although an interpretive framework is essential to prevent a theory from being observationally empty, it does not usually receive much attention, even though it must be imposed from the outside, never derived from within. The three theories considered here serve as an example of this. In a strong sense all three of them are really just one theory, on which three different schemes for interpretation have been imposed. In comparing their predictions with experiment, their quite different (and in the cases of RED and NED, incompletely developed) interpretive frameworks must be used side by side with their possibly different numerical results. Occasionally this has been forgotten.

ACKNOWLEDGMENTS

It is a pleasure to acknowledge discussions on the subject of these lectures with J. R. Ackerhalt, T. H. Boyer, S. J. Brodsky, C. K. Iddings, E. T. Jaynes, P. L. Knight, L. Mandel, P. W. Milonni, E. A. Power, M. O. Scully, C. R. Stroud, Jr., and K. Wódkiewicz.

References and Notes

1. By classical we mean that none of the interpretive framework of quantum field theory is involved, and that the dynamical fields are simple mathematical functions that commute everywhere and at all times with each other. Another discussion of possible consequences of (1) and (4) has been given by J. H. Eberly, in *Laser Photochemistry, Tunable Lasers, and Other Topics,* Eds. S. F. Jacobs *et al.*, Addison-Wesley, Reading, Massachussetts (1976), p. 421.

2. V. F. Weisskopf and E. Wigner, *Z. Phys.* **63**, 54 (1930). See also Section 28 of G. Källen, *Quantum Electrodynamics,* translated by C. K. Iddings and M. Mizushima, Springer, Heidelberg (1972).

3. J. R. Ackerhalt and J. H. Eberly, *Phys. Rev. D* **10**, 3350 (1974).

4. S. B. Lai, Ph.D. Thesis, University of Rochester (1976). See also P. Avan, C. Cohen-Tannoudji, J. Dupont-Roc, and C. Fabre, *J. Phys. (Paris)* **37**, 993 (1976); E. J. Moniz and D. H. Sharp, *Phys. Rev. D* **15**, 2850 (1976); and J. Dupont-Roc, C. Fabre, and C. Cohen-Tannoudji, *J. Phys. B* **11**, 563 (1978).

5. P. W. Milonni, *Phys. Rep.* **25**, 1 (1976); J. H. Eberly, reference 1; J. H. Eberly, in *Particles and Fields 1974,* Ed. C. E. Carlson, American Institute of Physics, New York (1974).

6. K. Wódkiewicz and J. H. Eberly, *Ann. Phys. (New York)* **101**, 574 (1976), and references therein.

7. P. W. Milonni and W. A. Smith, *Phys. Rev. A* **11**, 814 (1975); P. W. Milonni, J. R. Ackerhalt, and W. A. Smith, *Phys. Rev. Lett.* **31**, 958 (1973).

8. See T. H. Boyer, *Phys. Rev. D* **11**, 790, 809 (1975), and P. W. Milonni, reference 5.

9. See E. T. Jaynes, Microwave Laboratory Report No. 502, Stanford University, Stanford (1958).

10. E. T. Jaynes, in *Coherence and Quantum Optics,* Eds. L. Mandel and E. Wolf, Plenum, New York (1973), p. 35.

11. E. T. Jaynes, in *Quantum Electronics,* Ed. C. H. Townes, Columbia University Press, New York (1960), p. 288. The Hamiltonian form of our field theory has close parallels with this earlier development.

12. L. Mandel, in *Progress in Optics,* Vol. 13, Ed. E. Wolf, North-Holland, Amsterdam (1975). See also P. W. Milonni, ref. 5 above.

13. Our theory is obviously based directly on Jaynes' NCT. What we call NED may well be identical with NCT, but it should be Jaynes who says so.

14. J. R. Ackerhalt, J. H. Eberly, and P. L. Knight, in *Coherence and Quantum Optics,* Eds. L. Mandel and E. Wolf, Plenum, New York (1973), p. 635. See also the remarks of E. T. Jaynes, p. 68–69 of the same volume, and L. Allen and J. H. Eberly, *Optical Resonance and Two-Level Atoms,* Wiley, New York (1975), Section 7.4.

3

Quantum Beats

E. T. Jaynes

As noted before,[1] one of the most durable of all beliefs about QED is that it endows light with "granular" properties that somehow wipe out the interference effects of classical electromagnetic theory.

In 1954, Forrester, Gudmundsen, and Johnson[2] proposed to observe microwave beats between Zeeman components of a spectral line. The existence of the effect was promptly denied by some on grounds of quantum theory—as everybody knows, "a given photon interferes only with itself." Yet the photoelectric klystron worked, the beats were seen, and an important lesson was learned about the meaning and correct application of quantum theory.

But not for long. In 1956, Hanbury Brown and Twiss[3] proposed to measure stellar diameters by interference measurements involving fourth-order spatial correlation functions of the field. Again, quantum theorists denied the existence of the effect. I myself, at a seminar talk by E. M. Purcell, witnessed the reaction of one Nobel laureate who thought a hoax was being perpetrated. Yet the experiment worked—just as classical EM theory predicted—and again an important lesson was learned.

But not for long. With the appearance of the laser in the early 1960s, we were told that it was fundamentally impossible to observe beats between independently running lasers—a given photon interferes only with itself. Naturally, the beats appeared on schedule—just as classical EM theory predicted.

But this time Roy Glauber[4] produced a detailed theoretical analysis showing in great generality just *why* QED allows beats to be seen in photoelectric detection experiments, and what property of the field state vector determines this. For example, whether beats can or cannot be

E. T. Jaynes • Department of Physics, Washington University, St. Louis, Missouri 63130

seen in the photocurrent induced by the light is not determined by any two-point correlation function of the field (such as the expectation of the Poynting vector), because the state vector of the field may leave the phase of those beats completely indeterminate. Rather, the observable spectral density of the photocurrent is an instrumental constant times the Fourier transform of the fourth-order correlation function,

$$\phi(\tau) = \langle E^2(t)E^2(t + \tau)\rangle$$

at the photocathode. So the lesson was learned still another time and, thanks to Glauber, one might have thought it would now stay learned.

This lesson was learned, independently, at about the same time by many others. One of the most interesting documents recording this fact is the article of Gordon, Louisell, and Walker[5] on "Quantum Fluctuations and Noise in Parametric Processes." They state in the abstract:

> We find that a classical description of the input fields and of the amplification process is completely valid provided we take correctly into account the response of the amplifier to the input zero-point fields. This result is valid for inputs of arbitrarily small power.

Their evident surprise at finding this from a QED analysis is shown by the number of times essentially the same remark is repeated in the article. And indeed, their result conflicts with what we have all been told about QED in undergraduate physics courses.

One can then imagine my dismay on receiving, in February 1974, a paper coauthored by one of my own former students with the title "Missing Interference Effects in Multi-Level Fluorescence." Ten years after Glauber's explanation, this most durable of all views about QED had returned still another time. Of course, with each resurrection it had appeared in a more subtle form; this time it was vastly more subtle and even seemed, at first glance, to be backed up by a specific QED calculation. However, this work[6] calculated only the expectation of the Poynting vector in the radiation from a single atom, and was thus, on two counts, inadequate to determine what QED predicts for a real experiment to observe beats in the photocurrent.

Some of these limitations were recognized and removed by Chow, Scully, and Stoner,[7] and a still more realistic treatment was given by Senitzky,[8] whose conclusions are, I think, a reliable guide to what should and should not be observable in a real experiment according to QED.

But all these analyses still leave untouched the basic point of principle about the difference in physical content of QED and classical electromagnetic theory, as it pertains to interference. So some pedagogy is still needed if erroneous ideas about interference effects in QED are not to crop up still another time in the future.

Deferring a more complete analysis of the specific experiments that might now be attempted and what value they would have as further tests of QED, this short note tries to explain only that basic point of principle. In view of my previous strong criticisms[1,9] of the Copenhagen interpretation of quantum theory, I must now step rather far out of character and expound that interpretation as it applies to this problem. Fortunately, one need not believe a theory in order to understand it and teach it. Quite the contrary, I am convinced that many who defend the Copenhagen interpretation most fervently do so only because they have never thought deeply enough to realize its full implications.

An atom can decay from an excited state $|a\rangle$ to a lower level $|b\rangle$, with emission of a photon $\hbar\omega_\alpha$, or to a different lower level $|c\rangle$ by emission of a photon $\hbar\omega_\beta$ of a different frequency. The point at issue was stated originally as: if the atom is allowed to decay spontaneously, choosing its own way through these decay modes, can we then see "lower-level beats" between the two photons? But of course, if a single atom can produce at most only a single photoelectron, then no matter how it decays, we can hardly expect to see anything that could properly be called "beats." So let us frame the question less presumptively: "Is it possible to see interference effects between the two photons?"

In classical theory it is enough to ask merely: "Does the atom emit waves of both frequencies simultaneously?" If the answer is "yes," this is taken to be a statement of physical fact—that electromagnetic waves of frequencies ω_α, ω_β *exist* in the space around the atom. This being the case, there is no reason why an experimenter could not, by one means or another, verify this by causing them to induce oscillations of the beat frequency $\omega_\alpha - \omega_\beta$ in some detection device. But whether these beats can or cannot be seen is entirely a question of the skill and ingenuity of the experimenter and does not concern the theoretician, whose job is finished when he has described, by calculation, the *real physical situation*. Or to put it more strongly: in classical physics, it is simply not in his area of competence for the theoretician to make pronouncements about what can and cannot be measured in the laboratory, any more than for the experimenter to proclaim what can and cannot be calculated in the theory.

As both Bohr and Heisenberg have stressed strongly and repeatedly, the situation in quantum theory is entirely different. Present (Copenhagen) quantum theory cannot, as a matter of principle, answer any question of the type, "What is really happening when . . . ?" because there is no longer any such thing as a "real physical situation." It can answer only: "What will be the result of this particular experiment?" Even then it usually gives, not a definite answer, but a set of possible answers, with their probabilities. As Bohr stressed over and over again, a physical phenomenon is defined *only* by specifying the entire experimental arrangement used to observe it. (Bohr even added the stricture that the experimental arrangement must be described in classical terms, although Wigner[10] asked, plaintively, ''But *why* must I describe it in classical terms? What will happen to me if I don't?'')

Bohr's point still has to be stressed today, in spite of the fact that we have all heard it so many times, because of an irresistible tendency to accept it in principle—and immediately ignore it in practice. This is seen in all the aforementioned discussions of ''lower-level quantum beats.'' But it is not merely an observation to which one can give lip service and then, as practical people, proceed as if it had never been uttered. For in quantum theory you cannot carry out a practical calculation of a real experimental result unless you know what experimental arrangement is to be used.

In quantum theory, then, the theoretician and experimenter are forced into a much closer collaboration—in principle, the theoretician's problem cannot even be formulated until the experimenter tells him exactly what apparatus and procedure he proposes to use. Whether ''lower level quantum beats'' can or cannot be observed according to QED depends on what experiment you perform and how you perform it.

In the notation of Chow, Scully, and Stoner (hereafter referred to as CSS), an initial state $\Psi(0) = |a0\rangle$ (i.e., atom in state a, field in vacuum state) evolves with time into lower states with emission of two photons, $\hbar\omega_\alpha$, $\hbar\omega_\beta$;

$$\Psi(t) = A_0(t)|a0\rangle + A_1(t)|b1_\alpha\rangle + A_2(t)|c1_\beta\rangle \tag{1}$$

where $|b1_\alpha\rangle$ denotes an atom in state $|b\rangle$, one photon $\hbar\omega_\alpha$ in the field, etc. Suppose the two transitions are equally rapid: $|A_1| = |A_2|$. Then, after the emission is over, we may write the final state $\Psi(\infty)$ as

$$\Psi(\infty) = (|b1_\alpha\rangle + e^{i\theta}|c1_\beta\rangle)/2^{1/2} \tag{2}$$

where θ is a phase factor determined by the equations of motion. This is the correlated state.

If now a measurement finds the atom to be in state $|b\rangle$, then according to present quantum mechanics the ''reduction of the wave packet'' occurs, and we know that the field must be in state $|1_\alpha\rangle$. Then, as noted by CSS, there can be no interference between ω_α, ω_β for any subsequent experiment, because to put it colloquially, the photon $\hbar\omega_\beta$ was ''never emitted.'' More specifically, if F represents any field observable, its expectation is $\langle F\rangle = \langle 1_\alpha|F|1_\alpha\rangle$, and ω_β is not involved at all. On these grounds, it has been held that lower-state interferences do not occur in QED.

But we can write the final state $\Psi(\infty)$ equally well as

$$\Psi(\infty) = (|\Psi_+\rangle|\alpha\beta_+\rangle + |\Psi_-\rangle|\alpha\beta_-\rangle)/2^{1/2} \tag{3}$$

where the field state $|\alpha\beta_\pm\rangle$ and atom states $|\psi_\pm\rangle$ are

$$|\alpha\beta_\pm\rangle = (|1_\alpha\rangle \pm e^{i\phi}|1_\beta\rangle)/2^{1/2} \tag{4}$$

$$|\Psi_\pm\rangle = (|b\rangle \pm e^{i(\theta-\phi)}|c\rangle)/2^{1/2} \tag{5}$$

and ϕ is a phase that we may choose arbitrarily. If now measurement shows the atom to be in state $|\psi_+\rangle$, then we know the field must be in the linear combination

$$|\alpha\beta\rangle = (|1_\alpha\rangle + e^{i\phi}|1_\beta\rangle)/2^{1/2} \tag{6}$$

in which both photons are present after all!

Any field observable F then has expectation value

$$\langle F\rangle = \tfrac{1}{2}(\langle\alpha|F|\alpha\rangle + \langle\beta|F|\beta\rangle + e^{i\phi}\langle\alpha|F|\beta\rangle + e^{-i\phi}\langle\beta|F|\alpha\rangle) \tag{7}$$

in which contributions from both photons, and interference terms between them, are present. So, it seems a little hasty to conclude that QED does not predict lower-level interference effects.

But is the measurement of $|\psi_\pm\rangle$ really possible? If a determination of the states $|b\rangle$, $|c\rangle$, as assumed by CSS, is possible, then a measurement of $|\psi_\pm\rangle$ is surely also possible. In the Bloch sphere representation, $|\psi_\pm\rangle$ are points lying on the equator, diametrically opposite, with longitude determined by our choice of ϕ. Application of a suitable 90° pulse (via the quadrupole moment matrix element $\langle b|Q|c\rangle$, if the dipole moment vanishes), will then bring about the transformation $|\psi_+\rangle \to |b\rangle$, $|\psi_-\rangle \to |c\rangle$, after which we use the apparatus of CSS.

We have, then, the full EPR paradox—and more. By applying or not applying the 90° pulse before measuring the atomic state we can, at will, force the radiation field into either: (1) a state with a known one of the photons $\hbar\omega_\alpha$, $\hbar\omega_\beta$ present, and no possibility of interference effects in any subsequent measurement; (2) a state with both $\hbar\omega_\alpha$, $\hbar\omega_\beta$ present with a known relative phase. Lower-level interference effects are then not only observable, but predictable. And we can decide which to *do* after the emission is over and the radiation is far from the atom, so there can be no thought of any physical influence on the radiation!

But that is not the end, because this radiation can still be recaptured totally and used to initiate a second experiment. Place the emitting atom A at one focus of a perfectly reflecting ellipsoidal cavity of major diameter D. After a time D/c the radiation converges, in a spherical wave of just the original intensity, onto the other focus, where we have placed a test atom A'.

Let A' have the inverted level scheme, i.e., it waits in its ground state $|a'\rangle$ from which it can be excited to $|b'\rangle$ or $|c'\rangle$ by absorption of $\hbar\omega_\alpha$, $\hbar\omega_\beta$, respectively. Immediately after the radiation has fallen on A', we measure its state, and we can at our option either use or not use a 90° preparatory pulse on A', independently of whether we used it on A.

We can, of course, measure the state of atom A while the radiation is en route, and since the distance $A - A'$ is less than the distance D traveled by the light, the result of the measurement can be transmitted to an experimenter E' stationed at A', reaching him before the light from atom A does. So E' can still decide which experiment to perform on A' *after* he knows which measurement was made on A, and its result.

QED then predicts the following. If atom A was found in state $|b\rangle$, then we know in advance that A' can be found later in $|b'\rangle$, but not in $|c'\rangle$, and vice versa. If we use the 90° pulse and find A in $|\psi_+\rangle$, then we know in advance that a similar measurement on A' can find it in $|\psi'_+\rangle$. The observable and predictable interference effect is then that A' cannot be found in $|\psi'_-\rangle$.

And bear in mind that all this holds even though, by the analog of (5), the relative phase with which $|b'\rangle$ and $|c'\rangle$ are combined in this state $|\psi'_-\rangle$, which becomes impossible, can be chosen arbitrarily by us. That is, by choosing[1] fine details of how the 90° pulse is applied to atom A, *after* it has decayed, we can force atom A' into a known linear combination of states $|b'\rangle$ and $|c'\rangle$ with any relative phase we please, and *the results of subsequent experiments on A' depend on that phase.*

From this, it is pretty clear why present quantum theory not only does not use—it does not even dare to mention—the notion of a "real physical situation." Defenders of the theory say that this notion is philosophically naive, a throwback to outmoded ways of thinking, and that recognition of this constitutes deep new wisdom about the nature of human knowledge. I say that it constitutes a violent irrationality, that somewhere in this theory the distinction between reality and our knowledge of reality has become lost, and the result has more the character of medieval necromancy than of science. It has been my hope that quantum optics, with its vast new technological capability, might be able to provide the experimental clue that will show us how to resolve these contradictions.

References and Notes

1. E. T. Jaynes, in *Coherence and Quantum Optics*, Eds. L. Mandel and E. Wolf, Plenum, New York (1972); pp. 35-81. How the phase θ_0 of the pumping signal controls the phase ϕ of equation (5) above is shown in equations (61) and (63) of this reference and in more detail in C. R. Stroud and E. T. Jaynes, *Phys. Rev. A* **1**, 106 (1979); Sec. 4.

2. A. T. Forrester, R. A. Gudmundsen, and P. O. Johnson, *Phys. Rev.* **99,** 1691 (1955); A. T. Forrester, *Am. J. Phys.* **24,** 192 (1956).
3. R. Hanbury Brown and R. Q. Twiss, *Nature* **178,** 1447 (1956); and following paper by E. M. Purcell.
4. R. Glauber, in *Quantum Optics and Electronics,* Eds. C. DeWitt *et al.,* Gordon and Breach, New York (1964).
5. J. P. Gordon, W. H. Louisell, and L. R. Walker, *Phys. Rev.* **129,** 481 (1963).
6. R. Herman, H. Grotch, R. Kornblith, and J. H. Eberly, *Phys. Rev. A* **11,** 1389 (1975); I. C. Khoo and J. H. Eberly, *Phys. Rev. A* **14,** 2174 (1976).
7. W. W. Chow, M. O. Scully, and J. O. Stoner, Jr., *Phys. Rev. A* **11,** 1380 (1975).
8. I. R. Senitzky, *Phys. Rev. Lett.* **35,** 1755 (1975).
9. E. T. Jaynes, in *Coherence and Quantum Optics IV.* Eds. L. Mandel and E. Wolf, Plenum Press, New York (1978); pp. 495–510.
10. E. P. Wigner, Colloquium talk at Washington University, March 27, 1974.

On Quantum Beat Phenomena and the Internal Consistency of Semiclassical Radiation Theories

Marlan O. Scully

The past decade has seen many successful applications of semiclassical radiation theory. In these calculations matter is treated quantum mechanically while radiation is described according to Maxwell's equations. In fact, if one adds "vacuum fluctuations" to this semiclassical picture, it would seem that we have all that we need to understand everything from spontaneous emission and the Lamb shift to superfluorescence and the laser linewidth in a semiquantitative fashion.

Studies along these lines have provided many worthwhile exercises in rethinking the foundations of radiation theory. However, it is important to bear in mind that there are subtle questions concerning the internal consistency of any theory which quantizes only matter while treating the field classically. This point is made in the following quote from Heitler[1]:

> From this point of view it becomes clear that quantum electrodynamics and quantum mechanics form two inseparable parts of a single body of quantum theory and neither is consistent without the other.

It is the purpose of this note to emphasize that quantum beat[2] phenomena provide us with a simple example of a case in which the results of a self-consistent fully quantized (QED) calculation differ substantially from those obtained via a semiclassical theory (SCT) even when augmented by the notion of vacuum fluctuations. This is one of the few problems known to the author which cannot be "explained," let alone calculated, by vacuum-fluctuation-type arguments.

Having thus motivated our considerations let us now turn to a dis-

Marlan O. Scully ● Department of Physics and Optical Sciences Center, University of Arizona, Tucson, Arizona. This work is supported by the Air Force Office of Scientific Research (AFSC), United States Air Force, and the Army Research Office, United States Army.

cussion of quantum beats via QED and SCT. As depicted in Figure 1, an ensemble of atoms prepared in a coherent superposition of states is described by a state vector,

$$|\psi(t)\rangle = A\exp(-i\omega_a t)|\psi_a\rangle + B\exp(-i\omega_b t)|\psi_b\rangle + C\exp(-i\omega_c t)|\psi_c\rangle \quad (1)$$

Furthermore, if the nonvanishing dipole matrix elements are denoted by

type-I atoms	type-II atoms
$\mathscr{P}_{ac} = e\langle\psi_a|r|\psi_c\rangle$	$\mathscr{P}_{ac} = e\langle\psi_a|r|\psi_c\rangle$
$\mathscr{P}_{bc} = e\langle\psi_b|r|\psi_c\rangle$	$\mathscr{P}_{bc} = e\langle\psi_a|r|\psi_b\rangle$

where the atomic designation I and II is explained in Figure 1, then the state (1) implies that each atom contains two microscopic oscillating dipoles, that is,

type-I atoms

type-II atoms

$$P(t) = \mathscr{P}_{ac}(A^*C)\exp(-i\omega_1 t) \qquad P(t) = \mathscr{P}_{ac}(A^*C)\exp(-i\omega_1 t)$$
$$+ \text{ c.c.} \qquad\qquad\qquad + \text{ c.c.}$$

$$\qquad\qquad\qquad\qquad\qquad\qquad\qquad\qquad\qquad (2)$$

$$+ \mathscr{P}_{bc}(B^*C)\exp(-i\omega_2 t) \qquad + \mathscr{P}_{ab}(A^*B)\exp(-i\omega_2 t)$$
$$+ \text{ c.c.} \qquad\qquad\qquad + \text{ c.c.}$$

where $\omega_1 = \omega_a - \omega_c$ and $\omega_2 = \omega_b - \omega_c$. From a semiclassical perspective the field radiated will then be a sum of two terms

$$E = \epsilon_1\exp(-i\omega_1 t) + \epsilon_2\exp(-i\omega_2 t) \quad (3)$$

in an obvious notation. Hence it is clear that a square law detector contains an interference or ''beat note'' term,

$$|E|^2 = \epsilon_1^2 + \epsilon_2^2 + \epsilon_1^*\epsilon_2\exp[i(\omega_1 - \omega_2)t] + \text{ c.c.} \quad (4)$$

Such a beat note is frequently observed in beam–foil spectroscopy experiments.

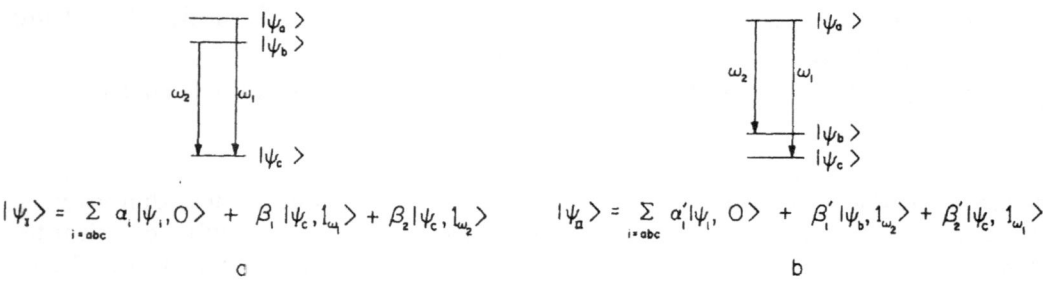

FIGURE 1

Finally we note, and this is the central point, that such an interference term is predicted by SCT for atoms of both types I and II.

Let us now consider the same problem as viewed from a QED perspective. For an atom of type I we now calculate a beat note

$$\langle \psi_I(t) | \hat{E}_1^\dagger(t) \hat{E}_2(t) | \psi_I(t) \rangle \tag{5}$$

where $\hat{E}_1^\dagger(t)$ and $\hat{E}_2(t)$ are proportional to the creation and annihilation operator expressions $\hat{a}_1^\dagger \exp(i\omega_2 t)$ and $\hat{a}_2 \exp(-i\omega_2 t)$, respectively. In view of $|\psi_I(t)\rangle$, as given in Figure 1a, expression (5) reduces to

$$\kappa \langle 1_{\omega_1} 0_{\omega_2} | a_1^\dagger a_2 | 0_{\omega_1} 1_{\omega_2} \rangle \exp[i(\omega_1 - \omega_2)t] \langle \psi_c | \psi_c \rangle$$

where κ is a constant. Hence, the beat note calculated via QED is given by

$$\kappa \exp[i(\omega_1 - \omega_2)t] \underbrace{\langle \psi_c | \psi_c \rangle}_{1} \tag{6}$$

On the other hand for type-II atoms we have

$$\langle \psi_{II}(t) | \hat{E}_1^\dagger(t) \hat{E}_2(t) | \psi_{II}(t) \rangle$$

and taking $|\psi_{II}\rangle$ from Figure 1b this becomes

$$
\begin{aligned}
&= \kappa' \langle 1_{\omega_1} 0_{\omega_2} | a_1^\dagger a_2 | 1_{\omega_2} 0_{\omega_2} \rangle \exp[i(\omega_1 - \omega_2)t] \langle \psi_c | \psi_b \rangle \\
&= \kappa' \exp[i(\omega_1 - \omega_2)t] \underbrace{\langle \psi_c | \psi_b \rangle}_{0}
\end{aligned}
\tag{7}
$$

Summarizing these QED considerations,

type-I atoms: $\langle \psi_I(t) | \hat{E}_1^\dagger(t) E_2(t) | \psi_I(t) \rangle = \kappa \exp[i(\omega_1 - \omega_2)t]$

type-II atoms: $\langle \psi_{II}(t) | \hat{E}_1^\dagger(t) E_2(t) | \psi_{II}(t) \rangle = 0$

whereas in the SCT calculations one finds the beat note amplitude to be nonvanishing for both type-I and type-II atoms.

The following argument based on the "quantum theory of measurement" provides some physical insight concerning the "missing" beats. A type-I atom when coherently excited will decay via the emission of a photon with frequency ω_1 or ω_2. Since both transitions lead to the same final atomic state, one cannot determine along which "path," ω_1 or ω_2, the atom decayed. Analogous to the Young's double-slit problem, this uncertainty in atomic trajectory leads to an interference between photons with frequencies ω_1 and ω_2, giving rise to quantum beats. A coherently excited type-II atom will also decay via the emission of a photon with frequency ω_1 or ω_2. However, after the emission is long past, an observation of the atom would now tell us which decay channel (1 or 2) was taken (atom in c or b). Consequently, we expect no beats in this case.

The clear conclusion is that a QED calculation is consistent with our most fundamental notions of quantum theory while SCT applied to this problem is not.

References and Notes

1. W. Heitler, *The Quantum Theory of Radiation*, Oxford University Press, New York (1956), p. 86.
2. A discussion of these ideas is given in Theory of coherent transients by M. O. Scully and R. F. Shea in *Laser Spectroscopy,* Eds. R. G. Brewer and A. Mooradian, Plenum Press, New York (1974), and a detailed account is given by W. Chow, M. Scully, and J. Stoner in *Phys. Rev.* **11**, 1380 (1975). See also, R. Herman, H. Grotch, Kronblith, and J. Eberly, *Phys. Rev.* **11**, 1389 (1975).

5

A Brief Survey of Stochastic Electrodynamics

Timothy H. Boyer

1. Introduction

Stochastic electrodynamics and random electrodynamics are the names given to a particular version of classical electrodynamics. This purely classical theory is Lorentz's classical electron theory[1] into which one introduces random electromagnetic radiation (classical zero-point radiation) as the boundary condition giving the homogeneous solution of Maxwell's equations. The theory contains one adjustable parameter setting the scale of the random radiation, and this parameter is chosen in terms of Planck's constant, $h = 2\pi\hbar$. Many of the researchers[2-70] working on stochastic electrodynamics hope that it will provide an accurate description of atomic physics and replace or explain quantum theory. At the very least the theory makes available new tools for calculating van der Waals forces, and it deepens our understanding of the connections between classical and quantum theories.[71]

The theory of stochastic electrodynamics has thus far been developed by a relatively small number of authors including Braffort,[2,3,8,9,11,12,42] Marshall,[4-8,10] Surdin,[9,43-49] Taroni,[9,11,12] Boyer,[13-35] Santos,[36-41,70] de la Pena-Auerbach and Cetto,[50-59] Theimer,[60-63] Peterson,[61-63] and Claverie and Diner.[64-67] It should be noted that there is some disagreement between these researchers as to just what has and has not been proved within the theory. It is the opinion of the present reviewer that a small but significant fraction of the articles published (and listed here in the references) on stochastic electrodynamics are exaggerated in their claims, and they are inaccurate in their reasoning and even in their elementary mathematics.

The reader should also be warned that this present review slights, indeed virtually ignores, one side of the research on stochastic electrodynamics, the detailed mathematical side, in favor of the side giving the

Timothy H. Boyer • Department of Physics, City College of the City University of New York, New York, New York 10031

physical results of the theory. One need only be reminded that early in this century two such excellent physicists as Einstein and von Laue had a controversy[72,73] in print regarding the stochastic properties of random classical radiation to realize that a rigorous mathematical treatment can be highly desirable. There is a growing literature[56,64–66,69,70] in stochastic electrodynamics that is involved with the investigation of the rigorous mathematical character of the Brownian motion of a charged particle in classical zero-point radiation.

2. Boundary Conditions in Classical Electron Theory

At the end of the nineteenth century and the beginning of the twentieth a classical theory of atomic structure was developed under the title of classical electron theory and under the preeminent leadership of Lorentz.[1] The theory consists of three basic assumptions: (*i*) The electromagnetic fields **E** and **B** satisfy Maxwell's equations with sources given by charged point masses; (*ii*) the charged point masses move according to Newton's second law under a force given by the Lorentz force $\mathbf{F} = e[\mathbf{E} + (\mathbf{v}/c) \times \mathbf{B}]$; (*iii*) the boundary conditions on the field differential equations correspond to the appearance of radiation only after the acceleration of particles.[74]

These assumptions appeared natural at the beginning of this century and still appear in every textbook of classical electrodynamics. Stochastic electrodynamics changes the third assumption. It is pointed out by the researchers on stochastic electrodynamics that the boundary conditions must be chosen so as to give the best possible description of nature. And one obtains a better description of nature if one introduces classical electromagnetic zero-point radiation as the boundary condition corresponding to the homogeneous solution of Maxwell's equations.

3. Connections with Traditional Classical and Quantum Theories

We will review shortly the assumed form of classical zero-point radiation. For the present we need only note that its energy spectrum $\rho(\omega)$ is given by

$$\rho(\omega) = (\omega^2/\pi^2 c^3)\,\tfrac{1}{2}\hbar\omega \tag{1}$$

which corresponds to an average energy $\tfrac{1}{2}\hbar\omega$ per normal mode of angular frequency ω. Planck's constant $h = 2\pi\hbar$ enters the theory as the parameter setting the scale of the zero-point radiation spectrum. Planck's constant is put into the theory at this one point and at no other. Every other appearance of Planck's constant is derived from its role as the scale factor in classical zero-point radiation.

Thus we see that stochastic electrodynamics is like traditional classical electron theory in having the same concepts of particle, force, and field, although it contains Planck's constant h. In the limit $h \to 0$, corresponding to a vanishing spectrum of zero-point radiation, we recover the boundary condition (*iii*) of traditional classical electron theory and hence recover exactly the familiar results of Lorentz's theory.

On the other hand, since stochastic electrodynamics contains Planck's constant h, it is in a sense similar to quantum electrodynamics. However, the stochastic theory is a classical theory and has none of the noncommuting operators on Hilbert space which appear in the quantum theory.

Thus stochastic electrodynamics stands in an intermediate position between traditional classical electron theory and classical electrodynamics. Exploration of stochastic electrodynamics should extend our ability to explain atomic phenomena in classical terms and it should clarify the limits of applicability of a purely classical theory.

4. Classical Electromagnetic Zero-Point Radiation

The idea of zero-point energy for mechanical systems arose in Planck's second theory[75] for the blackbody radiation spectrum in 1911. However, it seems to have been Nernst[76] who in 1916 extended the idea of zero-point energy to the free electromagnetic field giving electromagnetic zero-point radiation. Stimulated in part by a paper of Einstein and Stern[77] which assigned a zero-point energy $\hbar\omega_0$ to an oscillator, Nernst extended the idea, assigning a zero-point energy $\hbar\omega$ per normal mode to the radiation field. Nernst noted that this spectrum had the marvelous property of being invariant under an adiabatic compression[78] of a cavity with reflecting walls, but he was unsure whether the spectrum was Lorentz invariant.[79] Nernst introduced new postulates beyond classical theory and did not regard his work as simply a special version of classical electron theory. In 1948 Welton[80] considered classical zero-point radiation with energy $\frac{1}{2}\hbar\omega$ per normal mode to provide a semiclassical description of some quantum electrodynamic phenomena, while Casimir[81,82] calculated new results for van der Waals forces using quantum electromagnetic zero-point energy in a manner suggestive of classical theory. Braffort and Tzara[3] in 1954 reintroduced the zero-point radiation spectrum as a temperature-independent field in accord with Wien's law, $\rho(\omega) = \omega^3 F(\omega/T)$, and applied the spectrum as classical radiation ·to the stochastic behavior of a charged harmonic oscillator, noting that it gave an average oscillator energy in agreement with quantum theory. However, it was the careful and extensive investigations of Marshall[4−8,10] beginning in 1963 that initiated the present development of stochastic electrodynamics.

The introduction of classical zero-point radiation is the crucial step in converting traditional classical electron theory over to stochastic electrodynamics. The zero-point radiation spectrum (1) has a number of properties which make it uniquely suitable as the boundary condition giving the homogeneous solution of Maxwell's equations in stochastic electrodynamics. The zero-point radiation spectrum (1) is the unique spectrum of random classical electromagnetic radiation that is Lorentz invariant,[5,16] isotropic in every inertial frame,[26] invariant under an adiabatic compression,[76,18] and invariant under scattering by a dipole oscillator[26] moving with arbitrary constant velocity.

Zero-point radiation can be written as an expansion in plane waves:

$$\mathbf{E}_{zp}(\mathbf{r}, t) = \operatorname{Re} \sum_{\lambda=1}^{2} \int d^3k\, \hat{\boldsymbol{\epsilon}}(\mathbf{k}, \lambda) \mathfrak{h}_{zp}(\omega_k) \exp[i\mathbf{k}\cdot\mathbf{r} - i\omega t + i\theta(\mathbf{k}, \lambda)] \quad (2)$$

$$\mathbf{B}_{zp}(\mathbf{r}, t) = \operatorname{Re} \sum_{\lambda=1}^{2} \int d^3k\, \hat{\mathbf{k}} \times \hat{\boldsymbol{\epsilon}} \mathfrak{h}_{zp}(\omega_k) \exp[i\mathbf{k}\cdot\mathbf{r} - i\omega t + i\theta(\mathbf{k}, \lambda)] \quad (3)$$

where the polarization unit vectors $\hat{\boldsymbol{\epsilon}}(\mathbf{k}, \lambda)$ ($\lambda = 1, 2$) are orthogonal to the wave vector $\mathbf{k}$ and to each other, and

$$\pi^2 \mathfrak{h}_{zp}^2(\omega) = g_{zp}(\omega) = \tfrac{1}{2}\hbar\omega \quad (4)$$

corresponds to the electromagnetic energy per normal mode at frequency ω. The stochastic character[83] of the radiation is expressed in the random phase $\theta(\mathbf{k}, \lambda)$, which is distributed uniformly on $[0, 2\pi]$ and distributed independently for each $\mathbf{k}$ and λ.

5. Soluble Systems and Physical Problems

The presence of the random zero-point radiation leads to random zero-point fluctuations for all mechanical systems coupled to electromagnetic radiation. The behavior of the systems is described by stochastic differential equations where the stochastic character can be traced back to the random phases $\theta(\mathbf{k}, \lambda)$. The calculations seem quite difficult so that only a limited number of systems have been treated successfully at the present time. These systems include the free electromagnetic field,[5,27,40] the source-free electromagnetic field in the presence of conducting or dielectric boundaries,[6,10,13−15,19,25] the charged harmonic oscillator,[3,4,7,27,39,59] and the electric dipole rotator.[8,21,67] For the first three classes of systems one can establish general rules for the connections between classical and quantum theories. Even this limited set of soluble systems has been used to understand problems of physical interest including van der Waals forces,[6,10,14,15,19,22-28] diamagnetic behavior,[4,27,31,35] and radiation equilibrium.[16,17,29,30] There are tentative efforts to solve charged nonrelativistic nonlinear mechanical systems and relativistic mechanical systems. There have been extensive efforts to give a derivation of the Schrödinger equation from stochastic electrodynamics.[44,45,56,58]

6. Charged Nonrelativistic Harmonic Oscillator

The charged nonrelativistic harmonic oscillator is the perennial favorite of researchers in stochastic electrodynamics because it is simple enough to be solved in detail. Braffort and Tzara[3] considered this system in 1954 and Marshall[4] studied it extensively in 1963; it has been reexamined by Boyer,[27] Santos,[39] and de la Pena-Auerbach and Cetto.[59]

The system consists of a point charge e of mass m on the end of a spring of constant κ, giving a natural frequency $\omega_0 = (\kappa/m)^{1/2}$. The forces on the mass m are thus the restoring force of the spring $-m\omega_0^2 x$, the nonrelativistic damping force $2e^2\dddot{x}/3c^3$, and the force of the random radiation $eE_{\mathrm{zpx}}(0, t)$ taken in the dipole approximation:

$$m\ddot{x} = -m\omega_0^2 x + (2e^2/3c^3)\dddot{x} + eE_{\mathrm{zpx}}(0, t) \tag{5}$$

where $\mathbf{E}_{\mathrm{zp}}(\mathbf{r}, t)$ is the zero-point field given in (2).

This stochastic differential equation can be solved approximately (to order e^2 in the charge) to give the spreading of some arbitrary initial probability distribution on phase space. In the steady-state limit one finds a probability distribution[4] on phase space which is Gaussian in both displacement x and momentum $p = mv$,

$$P(x, p) = (\mathrm{const})\exp[-(p^2/2m + \tfrac{1}{2}m\omega_0^2 x^2)/\tfrac{1}{2}\hbar\omega_0] \tag{6}$$

From this distribution one can obtain the average energy of the system $\mathscr{E} = \tfrac{1}{2}\hbar\omega_0$, the mean-square displacement $\langle x^2 \rangle = \tfrac{1}{2}\hbar/m\omega_0$, and the mean-square momentum $\langle p^2 \rangle = \tfrac{1}{2}\hbar m\omega_0$. All of these results are identical with the expectation values obtained for a quantum oscillator in its ground state. Also we notice that the value of the charge e has disappeared from these lowest-order results so that they are in a sense universal. Indeed, by considering changes in the driving force and the radiation damping force, one can show that for any fixed charge distribution for the particle of mass m, the same results hold for $\mathscr{E}$, $\langle x^2 \rangle$, and $\langle p^2 \rangle$.

7. Some General Connections between Stochastic and Quantum Theories

The agreement between the stochastic and quantum results for the harmonic oscillator raises the question of the general areas of agreement and disagreement of the theories. For free electromagnetic fields and nonrelativistic harmonic oscillator systems it is possible to give the general connection.[27,38,40]

In the first place it is clear that the results of classical and quantum theories cannot be in complete agreement. Thus, the average value $\langle xp \rangle = \langle px \rangle = 0$ for the classical oscillator (5) in random classical radiation, whereas for the corresponding quantum oscillator the expectation value $\langle 0|\bar{x}\bar{p}|0 \rangle$ is not equal to $\langle 0|\bar{p}\bar{x}|0 \rangle$ in the ground state since the operators $\bar{x}$ and $\bar{p}$ do not commute; indeed $\langle 0|\bar{x}\bar{p}|0 \rangle = -\langle 0|\bar{p}\bar{x}|0 \rangle = \frac{1}{2}i\hbar$. In general, quantum theory has far more distinct combinations of $\bar{x}$ and $\bar{p}$ than the stochastic classical theory has of x and p precisely because the quantum operators do not commute whereas the classical terms do.

However, if one symmetrizes completely in the fundamental quantum variables, then the expectation value of the symmetrized quantum operator is the same as the average value of the corresponding classical expression in stochastic electrodynamics. Thus, in the example above, the symmetrized form $\frac{1}{2}(\bar{x}\bar{p} + \bar{p}\bar{x})$ has expectation value of $\langle 0|\frac{1}{2}(\bar{x}\bar{p} + \bar{p}\bar{x})|0 \rangle = 0$, in agreement with the corresponding classical average value $\langle xp \rangle = 0$. Moreover, one can show that this agreement holds at any temperature T if one chooses the Planck spectrum with zero-point radiation,

$$g(\omega, T) = \hbar\omega/[\exp(\hbar\omega/kT) - 1] + \tfrac{1}{2}\hbar\omega$$
$$= \tfrac{1}{2}\hbar\omega \coth(\hbar\omega/2kT) \tag{7}$$

as the boundary condition (4) for the homogeneous solution of Maxwell's equations.

Now this result reflects interest back on quantum theory because it singles out those quantum operators which are completely symmetrized in the fundamental quantum variables. The energy is always chosen as such a symmetrized operator. However, the operator for the angular momentum squared $\bar{\mathscr{L}}^2$ is not symmetrized and accordingly gives values which are different from those found in stochastic electrodynamics for a three-dimensional oscillator system. However, the value for $\bar{\mathscr{L}}^2$ used in the quantum literature differs from the symmetrized quantum operator by only the constant $\tfrac{3}{2}\hbar^2$ so that differences in the angular momentum squared remain the same between the various excited states of the quantum system. The use of symmetrized operators allows a much more understandable semiclassical description than that offered by the traditional quantum operator. For example, the traditional angular momentum operator requires that in the ground state of the hydrogen atom, where $\langle 0|\mathscr{L}^2|0\rangle = 0$, the electron must always be moving either directly towards the nucleus or directly away rather than ever moving around the nucleus.

8. Van der Waals Force Calculations

The first theoretical explanation of van der Waals forces between uncharged, unpolarized but polarizable materials was given in 1930 by London[84] in terms of quantum theory for two charged dipole oscillators. The semiclassical description of the quantum result involves spontaneous charge fluctuations which polarize the molecules. The semiclassical description must tantalize any physicist familiar with stochastic electrodynamics, where electromagnetic fluctuations appear due to classical zero-point radiation. Indeed, it turns out that stochastic electrodynamics can be applied successfully to the problem of van der Waals forces, giving results which are in agreement with quantum calculations when they exist, and providing classical calculations which are far simpler than those of quantum perturbation theory. One of the most successful applications of stochastic electrodynamics is in the calculation of van der Waals forces.

Casimir[81] in 1948 was the first to apply the ideas of electromagnetic zero-point energy to the van der Waals forces between macroscopic objects. He considered two conducting parallel plates and assigned an energy $\frac{1}{2}\hbar\omega$ for each classical normal mode allowed by the conducting boundary conditions. Extracting the separation-dependent part of the zero-point energy, he found an attractive force

$$F = -\pi^2\hbar c A/240d^4 \tag{8}$$

between good conductors of area A at a separation d. The force was measured experimentally by Sparnaay[85] in 1958 and has been termed one of the purest tests of quantum electrodynamics. However, exactly this same result appears in stochastic electrodynamics.[6,14,25] The calculation of van der Waals forces using stochastic electrodynamics can be extended to other geometries[14] and to dielectric[10] and permeable[25] materials. One merely uses the classical zero-point radiation fields (2), (3) and introduces the appropriate boundary conditions from classical electrodynamics. The forces can be calculated by finding the change in electromagnetic field energy when the materials are moved or by using the Maxwell stress tensor. The forces at finite temperature are easily calculated[6,10] in stochastic electrodynamics by inserting the appropriate Planck radiation expression (7) into the field expressions (2), (3), instead of inserting the zero-point spectrum (4).

In Casimir's conducting parallel plate calculation of 1948, all of the energy is stored in the electromagnetic field. Casimir[82] also calculated the force between a dipole oscillator and a conducting wall and the force between two electric dipole oscillators in the asymptotic limit of large distances where all the energy is again in the electromagnetic field. However, it was realized that within stochastic electrodynamics a complete calculation of the van der Waals forces between charged dipole oscillators at all distances could be carried out. The results[22–24] from stochastic electrodynamics are in exact agreement with those from quantum electrodynamics. Furthermore, the ease of calculation within classical theory allowed complete evaluation[28] of the forces at finite temperatures, something which has not been done in quantum theory. The proof[27] of the general connection between stochastic and quantum descriptions for free fields and harmonic oscillator systems assures us that here the much more difficult quantum calculations will indeed lead to the stochastic result.

9. Diamagnetism in Classical Theory

A qualitative understanding of diamagnetism can be given within classical theory even at the level of elementary physics.[86] If an atom is pictured as a point charge moving about a fixed center of force without radiating, then on turning on a magnetic field, the induced electric field speeds up or slows down the electron appropriately to give a diamagnetic effect.

Now this clear elementary result disappears on a more sophisticated analysis.[87] Bohr and van Leeuwen showed[88] in detail that if one applies classical statistical mechanics to the system, then all diamagnetic effects vanish. However, Bohr and van Leeuwen were working with traditional statistical mechanics; they did not include zero-point radiation.

It was Marshall[4] who first considered a three-dimensional charged harmonic oscillator in zero-point radiation and in the presence of a magnetic field. In the presence of zero-point radiation the Boltzmann statistical mechanics applied by Bohr and van Leeuwen do not hold and their null result collapses. Marshall found within stochastic electrodynamics a diamagnetic effect in precise agreement with the quantum result for all temperatures.[89] Landau's diamagnetism for a free charged particle has also been shown to exist within stochastic electrodynamics, again with a behavior in precise agreement with quantum theory.[35]

10. Efforts to Derive the Schrödinger Equation

It is hoped by the researchers in stochastic electrodynamics that it will be possible to understand quantum theory as some approximation to stochastic electrodynamics. Accordingly there has been considerable effort devoted to a derivation of the Schrödinger equation from the fluctuations of mechanical systems caused by classical zero-point radiation.

Of course there have been derivations of the Schrödinger equation from ideas of stochastic processes applied to the mechanics of classical particle systems.[90] However, within stochastic electrodynamics the analysis is allowed no freedom whatever. One cannot choose the stochastic process to fit the desired result; rather the stochastic behavior must follow from the random process introduced in the classical zero-point radiation forming the boundary condition on Maxwell's equations.

Several researchers[44,45,56,58] have claimed derivations of the Schrödinger equation within stochastic electrodynamics. However, none of these seems physically transparent, and the analyses have yet to provide a physical understanding of the stationary-state behavior which is an essential part of the Schrödinger analysis.

11. Connection between Stochastic Electrodynamics and the Old Quantum Theory

Stochastic electrodynamics and the old quantum theory both have roots in Lorentz's classical electron theory. Stochastic electrodynamics departs from Lorentz's theory in introducing classical zero-point radiation as a boundary condition. The old quantum theory ignores classical electromagnetic radiation but uses the mechanics of Lorentz's theory with the assumption that the system adiabatic invariants take on discrete values $J_i = \oint p_i dq_i = nh$, $n = 1, 2, 3, \ldots$. Now it turns out[31] that zero-point radiation is closely connected with adiabatic behavior and leads to average values for the adiabatic invariants given by $\langle J_i \rangle = \frac{1}{2}h$.

We have already remarked that zero-point radiation is the unique spectrum of radiation which is invariant[78,18] under adiabatic compression when in a closed container with reflecting walls. An adiabatic compression changes the energy of each mode and the frequency of each mode in just such a way that the zero-point radiation spectrum is invariant.

Also, if we consider a harmonic oscillator and change the frequency of the oscillator very slowly, then the ratio $\mathscr{E}/\omega$ of the oscillator energy $\mathscr{E}$ to frequency ω is an adiabatic invariant. For a charged oscillator interacting with random classical radiation, $\mathscr{E}/\omega$, where $\mathscr{E}$ is now the average energy, remains an adiabatic invariant[31] only when the random radiation spectrum is the zero-point spectrum (4).

Indeed, for the modes of the zero-point field itself and for charged mechanical systems which have no harmonics—such as the nonrelativistic harmonic oscillator, the rigid rotator, and the nonrelativistic three-dimensional harmonic oscillator in a magnetic field—it is possible to show[31] that the equilibrium distribution on phase space is given by

$$P(J_1, J_2, \ldots) = (\text{const})\exp[-2(\sum_i J_i)/h] \qquad (9)$$

where the J_i are the adiabatic invariants of the system. Moreover, the average $\langle J_i \rangle$ of the adiabatic invariant J_i is

$$\langle J_i \rangle = \tfrac{1}{2}h \qquad (10)$$

The appearance of Planck's constant in the phase space distribution arises from the fact that h is the scale factor in the zero-point radiation spectrum. For example, in the case of the one-dimensional oscillator leading to the phase space distribution (6) mentioned above, the adiabatic invariant is

$$J = \oint p\,dx = 2\pi(p^2/2m + \tfrac{1}{2}m\omega_0^2 x^2)/\omega_0 \qquad (11)$$

so that the distribution (6) is precisely of the form (9), $P(J) = (\text{const})\exp(-2J/h)$.

The use of adiabatic invariants, which was a hallmark of the old quantum theory, has a natural place within stochastic electrodsnamics. It is expected that the distribution in equation (9) will appear for a much wider class of systems, including relativistic systems, than has been treated up until now.

12. Qualitative Suggestions on Some Physical Problems

In addition to van der Waals forces and diamagnetic behavior, there are a number of other physical phenomena which seem strange and unexpected within traditional classical theory but are incorporated into quantum theory. These include the problem of atomic collapse, the Heisenberg uncertainty relations, the apparent wave nature of particles, the apparent particle nature of electromagnetic waves, the blackbody radiation spectrum, and sharp atomic spectra. Stochastic electrodynamics has not yet solved any of these additional puzzles within a classical context; however, there are tantalizing possibilities within the theory.

The most transparent case is that of atomic collapse. If we picture a hydrogen atom in traditional classical electron theory as a point electron moving around a massive nucleus, then the electron will accelerate, radiate, lose energy, and fall into the nucleus. However, if zero-point radiation is present as in stochastic electrodynamics, then the electron will absorb energy from the zero-point field and so there will arise a balance[8,26,67] between the absorption and emission of energy. The electron does not collapse into the nucleus. The scale of the atom will be determined by the scale of zero-point radiation since the amount of energy absorbed by the electron will depend upon the amount of random radiation which is present. Up until now only semiquantitative[26,67] or incomplete[8] calculations have been carried out for the hydrogen atom. However, precisely this same qualitative situation is seen for the harmonic oscillator mentioned earlier. There is a balance between energy emission and energy absorption which produces an equilibrium situation in which the oscillator has not lost all its energy, as it would in traditional classical theory. Rather the oscillator has an average energy $\langle \mathscr{E} \rangle = \frac{1}{2}\hbar\omega_0$ and fluctuating position $\langle x^2 \rangle = \frac{1}{2}\hbar/m\omega_0$, and momentum $\langle p^2 \rangle = \frac{1}{2}\hbar m\omega_0$, where the scale of the energy and the fluctuations are determined by the scale $\hbar$ of zero-point radiation. Thus stochastic electrodynamics suggests a qualitaive understanding of atomic size and of the avoidance of atomic collapse, but detailed calculations are so difficult that only the harmonic oscillator has been carried through in all detail.

The question of atomic spectra is a crucial but unresolved problem in stochastic electrodynamics. In the case of the harmonic oscillator the system involves only one frequency, that of the natural frequency of the oscillator, and agreement between the classical and quantum results is not a significant test.[7] In the case of the planar electric dipole rotator in classical zero-point radiation, Claverie and Diner [67] have pointed out that linear response theory gives a maximum in the net absorption of incident radiation at $\omega = \hbar/I$, which corresponds to the discrete line absorption frequency found for the quantum electric dipole rotator in its ground state. This result seems interesting since the classical rotator has a continuous distribution of frequencies of rotation in zero-point radiation.

The Heisenberg uncertainty relations are often interpreted semiclassically in terms of fluctuations. Of course, the zero-point radiation in stochastic electrodynamics implies fluctuations for all systems which have electromagnetic interactions, and for the harmonic oscillator in equilibrium considered above we indeed find that $\langle x^2 \rangle^{1/2}\langle p^2 \rangle^{1/2} = \frac{1}{2}\hbar$. It has been suggested[18] that the fluctuations of classical zero-point radiation, in addition to those of thermal radiation, may account for some of the fluctuation phenomena which are in quantum theory ascribed to photon statistics. Furthermore, one can at least conceive that the diffraction patterns observed when particles pass through gratings may arise from the interaction between the particles and the zero-point radiation where the pattern of zero-point radiation has been altered by the presence of the grating. Stochastic electrodynamics contains these qualitative possibilities, but only detailed calculation will reveal whether the theory does indeed describe the physical observations.

The problem of thermal radiation has been discussed within stochastic electrodynamics with some suggestion[16,18,60] that the experimentally observed Planck spectrum may be a result of the theory. However, the problem is not yet solved[29,30,61,63,68] and it seems crucially connected with the possibility of understanding atomic spectra.

The view taken within stochastic electrodynamics is that atomic systems are in equilibrium with zero-point radiation. The random radiation gives them their structure and indeed leads to van der Waals forces and diamagnetic behavior exactly as described earlier in this review. When an atomic system is excited so that it is not in equilibrium with zero-point radiation, then it emits radiation and this radiation above the zero-point radiation is detected by some laboratory measuring device. Now if stochastic electrodynamics is to give an acceptable description of atomic physics, we must indeed find that charged particle systems are in equilibrium with zero-point radiation, and we must explain why the radiation seen above the zero-point radiation appears in the form of relatively sharp spectral lines rather than in broad spectra with radiation at the frequency of mechanical oscillation.

Now the systems without harmonics—the nonrelativistic harmonic oscillator, the rotator, and the nonrelativistic three-dimensional harmonic oscillator in a magnetic field—when considered in the narrow linewidth approximation corresponding to weak coupling to the electromagnetic field, are all trivially in equilibrium with zero-point radiation.[30] Also the nonrelativistic harmonic oscillator, even when considered for finite linewidth, is in equilibrium with any isotropic distribution of radiation because it is linear.[26] However, if we go to charged nonrelativistic nonlinear systems, we find that the systems are not in equilibrium with zero-point radiation, but rather are in equilibrium with a radiation distribution that follows the Rayleigh–Jeans law.[29,30] A nonrelativistic nonlinear scattering system in zero-point radiation absorbs energy out of the higher radiation frequencies and radiates this energy away at low frequencies, effectively pushing the zero-point spectrum towards the Rayleigh–Jeans spectrum.[29,30] Such behavior seems intolerable if stochastic electrodynamics is to be a successful theory of atomic structure.

Now it has been conjectured[29,30] that relativity may play a crucial role in radiation equilibrium within stochastic electrodynamics. Very recently it has been proved[32,34] that the Rayleigh–Jeans law for thermal radiation is inconsistent with the Boltzmann distribution for relativistic free particles. If nonrelativistic mechanics is used, then indeed the Rayleigh–Jeans law and the Boltzmann distribution fit together neatly. However, special relativity for the mechanical system introduces a new element into the problem of radiation equilibrium. Clearly the use of relativistic mechanics, as opposed to nonrelativistic mechanics, is the appropriate choice for consistency with the relativistic content of Maxwell's equations. Thus the equilibrium distribution for random radiation within relativistic classical electrodynamics is unknown. Tied with the question of atomic radiation equilibrium is also the problem of radiation spectra. Only when we accurately understand the equilibrium situation of radiation emission and absorption can we hope to fully understand the radiation which exceeds the equilibrium value.

13. Conclusion

Over sixty years have elapsed since Nernst suggested the idea of zero-point radiation for the electromagnetic field, and well over a decade has passed since Marshall's striking and extensive work on stochastic electrodynamics during the early 1960s. There has been clarification of the theory, efforts to understand the general stochastic motion of charged particles in zero-point radiation, extensive calculation for systems of free fields and for harmonic oscillator systems, some connections shown between stochastic and quantum electrodynamics, and the start of a tie to the old quantum theory. However, major puzzles of quantum phenomena—thermal radiation equilibrium, line spectra, the apparent wave nature of particles—all lie close to but still outside the purview of stochastic electrodynamics. The success or failure of stochastic electrodynamics as a physical theory is still undecided.

References and Notes

1. H. A. Lorentz, *The Theory of Electrons*, Dover, New York (1952). This is a republication of the 2nd edition of 1915.
2. P. Braffort, M. Spighel, and C. Tzara, *C.R. Acad. Sci.* **239**, 157 (1954).
3. P. Braffort and C. Tzara, *C.R. Acad. Sci.* **239**, 1779 (1954).

4. T. W. Marshall, *Proc. R. Soc. London Ser. A* **276**, 475 (1963).

5. T. W. Marshall, *Proc. Cambridge Philos. Soc.* **61**, 537 (1965).

6. T. W. Marshall, *Nuovo Cimento* **38**, 206 (1965).

7. T. W. Marshall, *Izv. Vyssh. Uchelon. Zavedo Fiz.* **12**, 34, (1968).

8. T. W. Marshall and P. Claverie, *J. Math. Phys.*, to be published.

9. P. Braffort, M. Surdin, and A. Taroni, *C.R. Acad. Sci.* **261**, 4339 (1965).

10. L. L. Henry and T. W. Marshall, *Nuovo Cimento* **41**, 188 (1966).

11. M. Surdin, P. Braffort, and A. Taroni, *Nature* **210**, 405 (1966).

12. P. Braffort and A. Taroni, *C.R. Acad. Sci.* **264**, 1437 (1967).

13. T. H. Boyer, *Phys. Rev.* **174**, 1631 (1968).

14. T. H. Boyer, *Phys. Rev.* **174**, 1764 (1968).

15. T. H. Boyer, *Phys. Rev.* **180**, 19 (1968).

16. T. H. Boyer, *Phys. Rev.* **182**, 1374 (1969).

17. T. H. Boyer, *Phys. Rev.* **185**, 2039 (1969).

18. T. H. Boyer, *Phys. Rev.* **186**, 1304 (1969).

19. T. H. Boyer, *Ann. Phys. (N.Y.)* **56**, 474 (1970).

20. T. H. Boyer, *Phys. Rev. D* **1**, 1526 (1970).

21. T. H. Boyer, *Phys. Rev. D* **1**, 2257 (1970).

22. T. H. Boyer, *Phys. Rev. A* **5**, 1799 (1972).

23. T. H. Boyer, *Phys. Rev. A* **6**, 314 (1972).

24. T. H. Boyer, *Phys. Rev. A* **7**, 1832 (1973).

25. T. H. Boyer, *Phys. Rev. A* **9**, 2078 (1974).

26. T. H. Boyer, *Phys. Rev. D* **11**, 790 (1975).

27. T. H. Boyer, *Phys. Rev. D* **11**, 809 (1975).

28. T. H. Boyer, *Phys. Rev. A* **11**, 1650 (1975).

29. T. H. Boyer, *Phys. Rev. D* **13**, 2832 (1976).

30. T. H. Boyer, *Phys. Rev. A* **18**, 1228 (1978).

31. T. H. Boyer, *Phys. Rev. A* **18**, 1238 (1978).

32. T. H. Boyer, *Phys. Rev. D* **19**, 1112 (1979).

33. T. H. Boyer, *Phys. Rev. D* **19**, 3635 (1979).

34. T. H. Boyer, *Phys. Rev. A*, to be published.

35. T. H. Boyer, *Phys. Rev. A*, to be published.

36. E. Santos, *An. Re. Soc. Esp. Fis. Quim. Ser. A* **64**, 317 (1968).

37. E. Santos, *Lett. Nuovo Cimento* **4**, 497 (1972).

38. E. Santos, *J. Math. Phys. (N.Y.)* **15**, 1954 (1974).

39. E. Santos, *Nuovo Cimento B* **19**, 57 (1974).

40. E. Santos, *Nuovo Cimento B* **22**, 201 (1974).

41. E. Santos, *An. Fis.* **71**, 329 (1975).

42. P. Braffort, *C.R. Acad. Sci. Ser. B* **270**, 12 (1970).

43. M. Surdin, *C.R. Acad. Sci. Ser. B* **270**, 193 (1970).

44. M. Surdin, *Ann. Inst. Henri Poincaré* **15**, 203 (1971).

45. M. Surdin, *Int. J. Theor. Phys.* **4**, 117 (1971).

46. M. Surdin, *Int. J. Theor. Phys.* **8**, 183 (1973).

47. M. Surdin, *Int. J. Theor. Phys.* **9**, 185 (1974).

48. M. Surdin, *C.R. Acad. Sci. Ser. B* **280**, 337 (1975).

49. M. Surdin, *Phys. Lett. A* **58**, 370 (1976).

50. L. de la Pena-Auerbach and A. M. Cetto, *Nuovo Cimento B* **10**, 592 (1972).

51. L. de la Pena-Auerbach and A. M. Cetto, *Phys. Lett. A* **47**, 183 (1974).

52. L. de la Pena-Auerbach and A. M. Cetto, *Rev. Mex. Fis.* **23**, (1974).

53. L. de la Pena-Auerbach and A. M. Cetto, *Found. Phys.* **5**, 355 (1975).

54. L. de la Pena-Auerbach and A. M. Cetto, *Phys. Lett. A* **56**, 253 (1976).

55. L. de la Pena-Auerbach and A. M. Cetto, *Rev. Mex. Fix.* **25**, 1 (1976).

56. L. de la Pena-Auerbach and A. M. Cetto, *J. Math. Phys.* **18**, 1612 (1977).

57. L. de la Pena-Auerbach and A. M. Cetto, *Phys. Lett. A* **62**, 389 (1977).

58. L. de la Pena-Auerbach and A. M. Cetto, *Found. Phys.* **8**, 191 (1978).

59. L. de la Pena-Auerbach and A. M. Cetto, *J. Math. Phys.* **20**, 469 (1979).

60. O. Theimer, *Phys. Rev. D* **4**, 1597 (1971).

61. O. Theimer and P. R. Peterson, *Phys. Rev. D* **10**, 3962 (1974).

62. O. Theimer and P. R. Peterson, *Nuovo Cimento Lett.* **13**, 279 (1975).

63. O. Theimer and P. R. Peterson, *Phys. Rev. D* **14**, 656 (1976).

64. P. Claverie and S. Diner, *C.R. Acad. Sci. Ser. B* **280**, 1 (1975).

65. P. Claverie and S. Diner, in *Localization and Delocalization in Quantum Chemistry*, Vol. II, Eds. O. Chalvet, R. Daudel, S. Diner, and J. P. Malrieu, Reidel, Dordrecht, Holland (1976).

66. P. Claverie and S. Diner, *Ann. Fond. L. de Broglie* **1**, 73 (1976).

67. P. Claverie and S. Diner, *Int. J. Quantum Chem.* **12**, Suppl. 1, 41 (1977).

68. A. F. Kracklauer, *Phys. Rev. D* **14**, 654 (1976).

69. S. M. Moore, *Lett. Nuovo Cimento* **20**, 676 (1977).

70. L. Pesquera and E. Santos, *Lett. Nuovo Cimento* **20**, 308 (1977).

71. Several reviews of stochastic electrodynamics are already available. See References 26 and 64, and P. W. Milonni, *Phys. Rep.* **25**, 1 (1976), Section 5.

72. A. Einstein and L. Hopf, *Ann. Phys. (Leipzig)* **33**, 1105 (1910); A. Einstein, *Ann. Phys. (Leipzig)* **47**, 879 (1915).

73. M. von Laue, *Ann. Phys. (Leipzig)* **47**, 853 (1915).

74. In Reference 1, note 6 (p. 240) gives Lorentz's explicit assumption regarding the boundary conditions.

75. M. Planck, *Verh. Dtsch. Phys. Ges.* **13**, 138 (1911).

76. W. Nernst, *Verh. Dtsch. Phys. Ges.* **18**, 83 (1916).

77. A. Einstein and O. Stern, *Ann. Phys. (Leipzig)* **40**, 551 (1913).

78. See reference 76, pp. 89–90.

79. See reference 76, p. 116.

80. T. A. Welton, *Phys. Rev.* **74**, 1157 (1948).

81. H. B. G. Casimir, *K. Ned. Akad. Wet. Versl. Gewon. Vergad. Afd. Natuvrkd.* **51**, 793 (1948).

82. H. B. G. Casimir, *J. Chim. Phys. Phys. Chim. Biol.* **46**, 407 (1949).

83. The random character of the radiation is discussed in References 72 and 73, and by M. Planck, *Theory of Heat Radiation*, Dover, New York (1959). A modern treatment is that of S. O. Rice in *Selected Papers on Noise and Stochastic Processes*, Ed. N. Wax, Dover, New York (1954), p. 133.

84. F. London, *Z. Phys.* **63**, 245 (1930).

85. M. J. Sparnaay, *Physica (Utrecht)* **24**, 751 (1958).

86. E. M. Purcell, *Electricity and Magnetism*, McGraw-Hill, New York (1965), p. 370.

87. See, for example, L. Rosenfeld, *Theory of Electrons*, Dover, New York (1965), pp. 45–47.

88. See J. H. van Vleck, *The Theory of Electric and Magnetic Susceptibilities*, Oxford University Press, London (1932), p. 94; or D. C. Mattis, *The Theory of Magnetism*, Harper and Row, New York (1965), p. 21.

89. See also References 27 and 31 for alternative calculations of diamagnetism.

90. See, for example, E. Nelson, *Phys. Rev.* **150**, 1079 (1966).

6

A Canonical Transformation in Neoclassical Radiation Theory

E. A. Power and T. Thirunamachandran

1. Introduction

There have been recent attempts[1-4] to study radiation theory from a classical point of view. One of particular interest that has been investigated extensively is the formulation due to Jaynes and co-workers[5] called the neoclassical theory. This theory can be developed from the Hamiltonian viewpoint from which equations of motion can be derived. For example, Maxwell's equations form one set of these equations of motion where the driving currents are c-numbers formed by taking expectation values of the quantum operator currents. Much of the previous work has been devoted to examining the solutions of these equations. Here we discuss the formal Hamiltonian development with special emphasis on the freedom allowed by canonical transformations. In particular, we make explicit canonical transformations analogous to those used in quantum electrodynamics.[6-9] For example, one such transformation gives a flexibility in the choice of the field $\mathbf{E}$ or $\mathbf{D}$ for the canonical momentum. This corresponds to the transformation from minimal coupling to a Hamiltonian in multipolar form. Despite the classical nature of the Hamiltonian, a detailed analysis of the transformation shows that the Schrödinger character of the underlying dynamics requires closure relations and other sum rules that reflect basic quantum behavior.

E. A. Power • Department of Mathematics and *T. Thirunamachandran* • Department of Chemistry. University College London, London WC1 England

2. *The Hamiltonian in Neoclassical Theory*

In the neoclassical theory of Jaynes,[5] the Hamiltonian for the system is

$$H(p_n(\alpha), q_n(\alpha); \mathbf{\Pi}(\mathbf{r}), \mathbf{A}(\mathbf{r}))$$
$$= H_{\text{rad}}(\mathbf{\Pi}(\mathbf{r}), \mathbf{A}(\mathbf{r})) + H_{\text{atoms}}(p_n(\alpha), q_n(\alpha)) + H_{\text{int}} \quad (1)$$

where

$$H_{\text{rad}} = (1/8\pi) \int [16\pi^2 c^2 \mathbf{\Pi}^2(\mathbf{r}) + [\text{curl } \mathbf{A}(\mathbf{r})]^2\} \, d\mathbf{r} \quad (2)$$

$$= (1/8\pi) \int \{[\mathbf{E}^{\perp 2}(\mathbf{r}) + \mathbf{B}^2(\mathbf{r})] d\mathbf{r} \quad (3)$$

$$H_{\text{atom}}(\alpha) = \tfrac{1}{2} \sum_n [p_n^2(\alpha) + \omega_n^2(\alpha) q_n^2(\alpha)] \quad (4)$$

and

$$H_{\text{int}} = \langle V \rangle + \langle H_{\text{inter-Coulombic}} \rangle \quad (5)$$

In equations (1) and (2), $\mathbf{\Pi}(\mathbf{r})$ is the field conjugate to the vector potential $\mathbf{A}(\mathbf{r})$; in the minimal-coupling form assumed in equation (3), $\mathbf{\Pi}(\mathbf{r})$ is $-(1/4\pi c)\mathbf{E}^{\perp}(\mathbf{r})$ in terms of the transverse electric field. The Hamiltonian (4) for the atom is expressed in terms of classical canonical coordinates p_n, q_n (suppressing the index α) and underlying this classical description there is the quantum structure where the atom has spectrum $\hbar\omega_n$ with stationary states $\varphi_n(\mathbf{q})$. These canonical variables should not be confused with the usual quantum operators $\mathbf{p}$, $\mathbf{q}$ for the momentum and position variables. The p_n, q_n are defined through the projections a_n of an arbitrary state $\psi(\mathbf{q})$ onto the states $\varphi_n(\mathbf{q})$;

$$p_n = 2^{-1/2}(a_n + a_n^\dagger)(\hbar\omega_n)^{1/2} \quad (6)$$

$$q_n = 2^{-1/2}i(a_n - a_n^\dagger)(\hbar/\omega_n)^{1/2} \quad (7)$$

where

$$a_n = \int \bar{\varphi}_n(\mathbf{q})\psi(\mathbf{q}) d\mathbf{q} \quad (8)$$

The harmonic oscillator form of equation (4) has been discussed by Jaynes. It is analogous to the second-quantization procedure in quantum electrodynamics; but in the above Hamiltonian all variables are *c*-numbers. The term $\langle V \rangle$ in equation (5) is a function of the p_n's and q_n's. In the minimal-coupling framework,

$$\langle V \rangle = \sum_{m,n} a_n a_m^\dagger \langle \varphi_m(\mathbf{q}) \left| \frac{e\mathbf{p} \cdot \mathbf{A}(\mathbf{q})}{mc} + \frac{e^2}{2mc^2} \mathbf{A}^2(\mathbf{q}) \right| \varphi_n(\mathbf{q}) \rangle \tag{9}$$

$$\equiv \sum_{m,n} a_n a_m^\dagger V^{mn} \tag{10}$$

$$= \tfrac{1}{2} \sum_{m,n} \left(\frac{(p_n p_m + \omega_n \omega_m q_n q_m)}{(\hbar\omega_n)^{1/2}(\hbar\omega_m)^{1/2}} V^{(mn)} \right.$$

$$\left. + \frac{(p_n q_m \omega_m - q_m p_n \omega_n)}{(\hbar\omega_n)^{1/2}(\hbar\omega_m)^{1/2}} V^{[mn]} \right) \tag{11}$$

where $V^{(mn)}$ and $V^{[mn]}$ are the symmetric and antisymmetric parts of V^{mn} Since we are discussing a system with many atoms, the interaction term (5) also contains the inter-Coulombic interaction between charges on different atoms [the intra-Coulombic energies are naturally incorporated in (4)].[8]

The classical equations of motion implied by the Hamiltonian (1) are, for the radiation field,

$$\text{curl } \mathbf{B}(\mathbf{r}) = (1/c)(\partial \mathbf{E}^\perp / \partial t) + (4\pi/c)\mathbf{j}^\perp(\mathbf{r}) \tag{12}$$

and for the canonical variables $q_n(\alpha)$ and $p_n(\alpha)$ of the atom (α),

$$\dot{q}_n(\alpha) = p_n(\alpha) + \sum_m V^{(nm)} p_n(\alpha) + \sum_m V^{[nm]} \omega_m(\alpha) q_m(\alpha)$$

$$+ \text{ interatomic terms} \tag{13}$$

$$\dot{p}_n(\alpha) = -\omega_n^2(\alpha) q_n(\alpha) - \sum_m V^{[nm]} \omega_n(\alpha) p_m(\alpha)$$

$$- \sum_m V^{(nm)} \omega_n(\alpha) \omega_m(\alpha) q_m(\alpha) + \text{ interatomic terms} \tag{14}$$

The transverse current $\mathbf{j}^\perp(\mathbf{r})$ in equation (12) is

$$\mathbf{j}^\perp(\mathbf{r}) = \tfrac{1}{2} \sum_{m,n} \frac{(p_n p_m + q_n q_m \omega_n \omega_m)}{(\hbar\omega_n \hbar\omega_m)^{1/2}} \mathbf{j}^{(mn)}(\mathbf{r})$$

$$+ \frac{i}{2} \sum_{m,n} \frac{(p_n q_m \omega_m - p_m q_n \omega_n)}{(\hbar\omega_n \hbar\omega_m)^{1/2}} \mathbf{j}^{[mn]}(\mathbf{r}) \tag{15}$$

where

$$\mathbf{j}^{\perp mn}(\mathbf{r}) = \langle \varphi_m | (-e/2m)[\mathbf{p} + (e/c)\mathbf{A}(\mathbf{q})] \delta(\mathbf{r} - \mathbf{q}) | \varphi_n \rangle \tag{16}$$

The interatomic terms in equations (13) and (14) arise from the inter-Coulombic interaction term in equation (5). They are linear in the mode coordinates of the atom α, but quadratic in the mode coordinates of the other atoms.

3. Transformation to Multipolar Form

In this section we carry out a canonical transformation on (1) to obtain a new Hamiltonian which is the classical analogue of the quantum-mechanical multipolar Hamiltonian. The classical canonical transformation required is a simple linear transformation in phase space for the canonical coordinates p_n, q_n, together with a translation of the field $\Pi(\mathbf{r})$, with $\mathbf{A}(\mathbf{r})$ unchanged. It is convenient to define the coefficients for the p_n, q_n transformation in terms of a single matrix $C_{nk}(\mathbf{A})$. We use

$$p_n^{\text{old}} = \tfrac{1}{2} \sum_k (\omega_n/\omega_k)^{1/2} [(C_{nk} + \bar{C}_{nk}) p_k^{\text{new}} - i\omega_k(C_{nk} - \bar{C}_{nk}) q_k^{\text{new}}] \tag{17}$$

$$q_n^{\text{old}} = \tfrac{1}{2} i \sum_k (\omega_n\omega_k)^{-1/2} [(C_{nk} - \bar{C}_{nk}) p_k^{\text{new}} - i\omega_k(C_{nk} + \bar{C}_{nk}) q_k^{\text{new}}] \tag{18}$$

The transformation of the canonical momentum field (transverse) $\Pi(\mathbf{r})$ is

$$\Pi^{\text{old}}(\mathbf{r}) = \Pi^{\text{new}}(\mathbf{r}) + \mathbf{P}^\perp(\mathbf{r})/c \tag{19}$$

where $\mathbf{P}(\mathbf{r})$ is the electric polarization field. For a given atom centered at $\mathbf{R}$, $\mathbf{P}(\mathbf{r})$ can be written in a multipole series as

$$P_i(r) = \sum_{m,n} [\mu_i^{nm}\delta(\mathbf{r} - \mathbf{R}) - Q_{ij}^{nm}\nabla_j\delta(\mathbf{r} - \mathbf{R}) + \cdots] a_n a_m^\dagger$$

$$= \tfrac{1}{2} \sum_{m,n} P_i^{(nm)}(\mathbf{r}) \frac{p_n p_m + \omega_n\omega_m q_n q_m}{(\hbar\omega_n\hbar\omega_m)^{1/2}} \tag{20}$$

where $\mathbf{P}_i^{(nm)}(\mathbf{r})$ is symmetric and given by

$$P_i^{(nm)}(\mathbf{r}) = P_i^{(nm)}\delta(\mathbf{r} - \mathbf{R})$$

$$= (\mu_i^{nm} - Q_{ij}^{nm}\nabla_j + \cdots)\delta(\mathbf{r} - \mathbf{R}) \tag{21}$$

The canonical nature of the transformation requires that the Poisson brackets

$$\{q_n^{\text{old}}, p_m^{\text{old}}\} = \{q_n^{\text{new}}, p_m^{\text{new}}\} = \delta_{nm} \tag{22}$$

$$\{A_i^{\text{old}}(\mathbf{r}'), \Pi_j^{\text{old}}(\mathbf{r}'')\} = \{A_i^{\text{new}}(\mathbf{r}'), \Pi_j^{\text{new}}(\mathbf{r}'')\} = \delta_{ij}^\perp(\mathbf{r}' - \mathbf{r}'') \tag{23}$$

and the eight Poisson brackets of all other pairs of variables vanish.

Equation (22) implies

$$\sum_k \left(\frac{\partial q_n^{old}}{\partial q_k^{new}} \frac{\partial p_m^{old}}{\partial p_k^{new}} - \frac{\partial q_n^{old}}{\partial p_k^{new}} \frac{\partial p_m^{old}}{\partial q_k^{new}} \right)$$

$$+ \sum_{l=1,2,3} \int \left(\frac{\partial q_n^{old}}{\partial A_l^{new}(\mathbf{r})} \frac{\partial p_m^{old}}{\partial \Pi_l^{new}(\mathbf{r})} - \frac{\partial q_n^{old}}{\partial \Pi_l^{new}(\mathbf{r})} \frac{\partial p_m^{old}}{\partial A_l^{new}(\mathbf{r})} \right) d\mathbf{r} \qquad (24)$$

$$= \sum_k \{ [(C_{nk} + \bar{C}_{nk})(C_{mk} + \bar{C}_{mk})/4] - [(C_{nk} - \bar{C}_{nk})(C_{mk} - \bar{C}_{mk})/4] \}$$

$$= \sum_k C_{nk} \bar{C}_{mk} = \delta_{nm} \qquad (25)$$

The integral term on the left-hand side of (24) is zero since neither p_m^{old} nor q_n^{old} depends on $\mathbf{\Pi}(\mathbf{r})$. Thus, equation (22) will be satisfied if $C_{nk}(\mathbf{A})$ is a unitary matrix. The Poisson bracket (23) involving the conjugate fields is automatically satisfied since $\mathbf{A}(\mathbf{r})$ is independent of p_k and q_k. Of the remaining Poisson brackets which are zero,

$$\{ p_n^{old}, \Pi_l^{old} \} = \{ p_n^{new}, \Pi_l^{new} \} = 0 \qquad (26)$$

gives rise to an important relationship for the matrix C_{nk}. We have

$$\sum_k \left(\frac{\partial p_n^{old}}{\partial q_k^{new}} \frac{\partial \Pi_l^{old}(\mathbf{r}')}{\partial p_k^{new}} - \frac{\partial p_n^{old}}{\partial p_k^{new}} \frac{\partial \Pi_l^{old}(\mathbf{r}')}{\partial q_k^{new}} \right)$$

$$+ \sum_{l=1,2,3} \int \left(\frac{\partial q_n^{old}}{\partial A_l^{new}(\mathbf{r})} \frac{\partial \Pi_l^{old}(\mathbf{r}')}{\partial \Pi_l^{new}(\mathbf{r})} - \frac{\partial p_n^{old}}{\partial \Pi_l^{new}(\mathbf{r})} \frac{\partial \Pi_l^{old}(\mathbf{r}')}{\partial A_l^{new}(\mathbf{r})} \right)$$

$$= \sum_k \frac{1}{2} \left(\frac{\omega_n}{\omega_k} \right)^{1/2} \left(\frac{\partial (C_{nk} + \bar{C}_{nk})}{\partial A_i(\mathbf{r}')} \right.$$

$$\left. - \frac{i}{\hbar c} \sum_m (C_{nm} - \bar{C}_{nm}) P_i^{mk} \delta(\mathbf{r}' - \mathbf{R}) \right) p_k$$

$$+ \sum_k - \frac{i}{2} (\omega_n \omega_k)^{1/2} \left(\frac{\partial (C_{nk} - \bar{C}_{nk})}{\partial A_i(\mathbf{r}')} \right.$$

$$\left. - \frac{i}{\hbar c} \sum_m (C_{nm} + \bar{C}_{nm}) P_i^{mk} \delta(\mathbf{r}' - \mathbf{R}) \right) q_k = 0 \qquad (27)$$

For this to be identically true, the coefficients of p_k and q_k must separately vanish. Clearly this requires

$$\partial C_{nk}/\partial A_i(\mathbf{r}') = \frac{i}{\hbar c} \sum_m C_{nm} P_i^{mk} \delta(\mathbf{r}' - \mathbf{R}) \qquad (28)$$

Thus,

$$C_{nk} = \{ \exp[(i/\hbar c) P_i A_i(\mathbf{R})] \}^{nk} \qquad (29)$$

We draw attention to the fact that C_{nk} is the matrix element of the exponential and not the exponential of P_i^{nk}. It may be added that this matrix, which gives the coefficients of the *classical* canonical transformations (17) and (18), has the same structure as the generator for the corresponding transformation in quantum electrodynamics.

The canonically transformed Hamiltonian is obtained from equation (1) by substitution of equations (17), (18), and (19). Using C_{nk} and the polarization field (20), the radiation Hamiltonian (2) transforms to

$$(1/8\pi) \int \{16\pi^2 c^2 [\Pi(\mathbf{r}) + \mathbf{P}^\perp(\mathbf{r})/c]^2 + \mathbf{B}^2(\mathbf{r})\} d\mathbf{r}$$

$$= H_{\mathrm{rad}} + \int 4\pi c \Pi(\mathbf{r}) \cdot \mathbf{P}^\perp(\mathbf{r}) d\mathbf{r} + 2\pi \int |\mathbf{P}^\perp(\mathbf{r})|^2 d\mathbf{r} \tag{30}$$

where $\Pi(\mathbf{r})$ is the new conjugate field and is now identified with $-(1/4\pi c)\mathbf{D}^\perp(\mathbf{r})$ according to the relation

$$\dot{\mathbf{A}} = -c\mathbf{E}^\perp = \partial H/\partial \Pi = 4\pi c^2 \Pi + 4\pi c \mathbf{P}^\perp \tag{31}$$

where

$$\mathbf{D} = \mathbf{E} + 4\pi \mathbf{P} \tag{32}$$

In fact, the right-hand side of (30) can be written as

$$H_{\mathrm{rad}} + H_{\mathrm{pol}} + H_{|\mathbf{P}^\perp|^2} \tag{33}$$

where

$$H_{\mathrm{pol}} = -\int \mathbf{D}^\perp(\mathbf{r}) \cdot \mathbf{P}(\mathbf{r}) d\mathbf{r} \tag{34}$$

The transformation of $H_{\mathrm{atom}} + V$, equations (4) and (11), gives

$$\frac{1}{2} \sum_{m,n} \{(p_n p_m + \omega_n \omega_m q_n q_m)[\delta_{nm} + (\hbar\omega_n \hbar\omega_m)^{-1/2} V^{(nm)}]$$

$$+ (p_n q_m \omega_m - q_n p_m \omega_n)(\hbar\omega_n \hbar\omega_m)^{-1/2} V^{[nm]}\}$$

$$\Rightarrow \frac{1}{2} \sum_{m,n,k,l} \left\{ \frac{(p_k p_l + \omega_k \omega_l q_k q_l)}{(\hbar\omega_k \hbar\omega_l)^{1/2}} \frac{1}{2} [C_{nk}(\hbar\omega_n \delta_{nm} + V^{(nm)})\bar{C}_{ml} \right.$$

$$+ \bar{C}_{nk}(\hbar\omega_n \delta_{nm} + V^{(nm)})C_{ml} + \bar{C}_{nk} V^{[nm]} C_{ml} - \bar{C}_{nk} V^{[nm]} \bar{C}_{ml}]$$

$$+ \frac{(p_k q_l \omega_l - p_l q_k \omega_k)}{(\hbar\omega_k \hbar\omega_l)^{1/2}} \frac{1}{2} i [C_{nk}(\hbar\omega_n \delta_{nm} + V^{(nm)})\bar{C}_{ml}$$

$$\left. - \bar{C}_{nk}(\hbar\omega_n \delta_{nm} + V^{(nm)})C_{ml} - \bar{C}_{nk} V^{[nm]} C_{ml} - C_{nk} V^{[nm]} \bar{C}_{ml}] \right\} \tag{35}$$

The term antisymmetric in k, l in equation (35) has the factor

$$\frac{1}{2} i [-\bar{C}_{kn}(\Delta + V)_{nm} C_{ml} + C_{kn}(\Delta + V)_{nm} \bar{C}_{ml} - 2 C_{nk} V^{[nm]} \bar{C}_{ml}]$$

$$= \frac{1}{2} i (-\bar{C}HC + CH\bar{C} - 2CV^{[1]}\bar{C})^{[kl]} \tag{36}$$

where

$$H = \Delta + V \quad \text{and} \quad \Delta_{nm} = \hbar\omega_n\delta_{nm} \tag{37}$$

If the matrix (29) is used to evaluate (36), we find that the factor is

$$-i\left\{ \int [P_i(\mathbf{r}), H]A_i(\mathbf{R})d\mathbf{r} + V^{[]} \right\}^{[kl]} \tag{38}$$

In the Appendix we show that

$$(-i/\hbar)[P(\mathbf{r}), H] = \mathbf{j}^{\perp}(\mathbf{r}) - c \operatorname{curl} \mathbf{M}(\mathbf{r}) \tag{39}$$

where $\mathbf{M}(\mathbf{r})$ is the magnetization field whose leading term in a multipole expansion is the well-known magnetic dipole, $-(e/mc)(\mathbf{q} - \mathbf{R}) \times \mathbf{p}\delta(\mathbf{r} - \mathbf{R})$; if this is inserted into (38), we have

$$-i\left\{ (1/c)\int \mathbf{j}^{\perp}(\mathbf{r})\cdot\mathbf{A}(\mathbf{r})d\mathbf{r} - \int [\operatorname{curl}\mathbf{M}(\mathbf{r})]\cdot\mathbf{A}(\mathbf{r})d\mathbf{r} + V^{[]} \right\}^{[kl]}$$

$$= i\left\{ \int [\operatorname{curl}\mathbf{M}(\mathbf{r})]\cdot\mathbf{A}(\mathbf{r})d\mathbf{r} \right\}^{[kl]} \tag{40}$$

$$= i\int \mathbf{M}^{[kl]}(\mathbf{r})\cdot\mathbf{B}(\mathbf{r})d\mathbf{r}$$

where we have made use of the obvious equality

$$(1/c)\int \mathbf{j}^{\perp[kl]}(\mathbf{r})\cdot\mathbf{A}(\mathbf{r})d\mathbf{r} = -V^{[kl]} \tag{41}$$

Thus the second term in equation (35) is

$$\tfrac{1}{2}\sum_{k,l} \frac{(p_k q_l \omega_l - p_l q_k \omega_k)}{(\hbar\omega_k\hbar\omega_l)^{1/2}} \, i\int \mathbf{M}^{[kl]}(\mathbf{r})\cdot\mathbf{B}(\mathbf{r})d\mathbf{r} \tag{42}$$

The symmetric term in the transformed Hamiltonian is obtained by a similar development. The detailed calculation is considerably more complicated because the result contains a term that is quadratic in the electric charge and in the magnetic field. The argument parallels that of reference 8; we obtain

$$\tfrac{1}{2}\sum_{k,l} \frac{(p_k p_l + \omega_k\omega_l q_k q_l)}{(\hbar\omega_k\hbar\omega_l)^{1/2}} [\hbar\omega_k\delta_{kl}$$

$$+ \tfrac{1}{2}\int O_{ij}^{(kl)}(\mathbf{r}, \mathbf{r}')B_i(\mathbf{r})B_j(\mathbf{r}')d\mathbf{r}d\mathbf{r}'] \tag{43}$$

where the leading term of the diamagnetic interaction, $\tfrac{1}{2}\int \mathbf{O}{:}\mathbf{BB} \, d\mathbf{r} \, d\mathbf{r}'$, is

$$(e^2/8mc^2)[(\mathbf{q} - \mathbf{R}) \times \mathbf{B}(\mathbf{R})]^2 \tag{44}$$

The new Hamiltonian in neoclassical framework can be written down from equations (33), (42), and (43):

$$H_{\text{multipolar}}\big(p_n(\alpha),\, q_n(\alpha);\, \mathbf{\Pi}(\mathbf{r}),\, \mathbf{A}(\mathbf{r})\big)$$

$$= H_{\text{rad}}\,\big(\mathbf{\Pi}(\mathbf{r}),\, \mathbf{A}(\mathbf{r})\big) + H_{\text{atoms}}\big(p_n(\alpha),\, q_n(\alpha)\big)$$

$$-\tfrac{1}{2}\sum_{k,l,\alpha} \frac{p_k(\alpha)p_l(\alpha) + \omega_k(\alpha)\omega_l(\alpha)q_k(\alpha)q_l(\alpha)}{[\hbar\omega_k(\alpha)\hbar\omega_l(\alpha)]^{1/2}}$$

$$\times \int \mathbf{P}^{(kl)}(\mathbf{r})\cdot\mathbf{D}^{\perp}(\mathbf{r})\,d\mathbf{r}$$

$$+\tfrac{1}{2}\sum_{k,l,\alpha} \frac{p_k(\alpha)q_l(\alpha)\omega_l(\alpha) - p_l(\alpha)q_k(\alpha)\omega_k(\alpha)}{[\hbar\omega_k(\alpha)\hbar\omega_l(\alpha)]^{1/2}}$$

$$\times\, i \int \mathbf{M}^{[kl]}(\mathbf{r})\cdot\mathbf{B}(\mathbf{r})\,d\mathbf{r}$$

$$+\tfrac{1}{2}\sum_{k,l,\alpha} \frac{p_k(\alpha)p_l(\alpha) + \omega_k(\alpha)\omega_l(\alpha)q_k(\alpha)q_l(\alpha)}{[\hbar\omega_k(\alpha)\langle\omega_l(\alpha)]^{1/2}}$$

$$\times \tfrac{1}{2}\int O^{(kl)}_{ih}(\mathbf{r},\mathbf{r}')B_i(\mathbf{r})B_j(\mathbf{r}')\,d\mathbf{r}\,d\mathbf{r}' \qquad (45)$$

In (45) we have omitted the self-energy terms that arise from the third term $2\pi \int |\mathbf{P}^{\perp}(\mathbf{r})|^2 d\mathbf{r}$, in equation (30). Equation (45) is the neoclassical analogue of the quantum-electrodynamical Hamiltonian in multipolar form.[8,9] The electric dipole approximation simplifies (45) to

$$H_{\text{elec. dip.}} = H_{\text{rad}} + H_{\text{atoms}}$$

$$-\tfrac{1}{2}\sum_{k,l,\alpha} \frac{p_k(\alpha)p_l(\alpha) + \omega_k(\alpha)\omega_l(\alpha)q_k(\alpha)q_l(\alpha)}{[\hbar\omega_k(\alpha)\hbar\omega_l(\alpha)]^{1/2}} \times \boldsymbol{\mu}^{kl}(\alpha)\cdot\mathbf{D}^{\perp}(\mathbf{R}_\alpha) \quad (46)$$

This may be compared with the dipole approximation to the original Hamiltonian (1) for which we have

$$H_{\text{elec. dip.}} = H_{\text{rad}} + H_{\text{atoms}}$$

$$+\tfrac{1}{2}\sum_{k,l,\alpha} \frac{p_l(\alpha)q_k(\alpha)\omega_k(\alpha) - q_l(\alpha)p_k(\alpha)\omega_l(\alpha)}{[\hbar\omega_k(\alpha)\hbar\omega_l(\alpha)]^{1/2}}$$

$$\times\, i\left(\frac{e}{mc}\right)\mathbf{p}^{kl}(\alpha)\cdot\mathbf{A}(\mathbf{R}_\alpha)$$

$$+\sum_{k,l,\alpha} \frac{p_k(\alpha)p_l(\alpha) + \omega_k(\alpha)\omega_l(\alpha)q_k(\alpha)q_l(\alpha)}{[\hbar\omega_k(\alpha)\hbar\omega_l(\alpha)]^{1/2}}$$

$$\times\left(\frac{e^2}{2mc^2}\right)\delta_{kl}\mathbf{A}^2(\mathbf{R}_\alpha)$$

$$+ H_{\text{inter-Coulombic}} \qquad (47)$$

We note that in contrast to (47), equation (46) does not explicitly contain any inter-Coulombic terms. All inter-atomic interactions are mediated by the transverse field. This is a general feature of the multipolar Hamiltonian (45).

4. Equations of Motion for the Multipolar Hamiltonian

For the mode coordinates q_n, p_n, the classical equations of motion resulting from the multipolar Hamiltonian are

$$
\dot{q}_n = \frac{\partial H_{\text{multi}}}{\partial p_n} = p_n - \sum_m \frac{p_m}{[\hbar\omega_m\hbar\omega_n]^{1/2}} \left[\int \mathbf{P}^{(nm)}(\mathbf{r}) \cdot \mathbf{D}^{\perp}(\mathbf{r}) \, d\mathbf{r} \right.
$$
$$
\left. - \tfrac{1}{2} \int O_{ij}^{(nm)}(\mathbf{r}, \mathbf{r}') B_i(\mathbf{r}) B_j(\mathbf{r}') \, d\mathbf{r} \, d\mathbf{r}' \right]
$$
$$
+ \sum_m \frac{\omega_m q_m}{[\hbar\omega_n\hbar\omega_m]^{1/2}} \, i \int \mathbf{M}^{[nm]}(\mathbf{r}) \cdot \mathbf{B}(\mathbf{r}) \, d\mathbf{r} \tag{48}
$$

and

$$
\dot{p}_n = -\frac{\partial H_{\text{multi}}}{\partial q_n} = -\omega_n^2 q_n + \sum_m \frac{\omega_n\omega_m q_m}{[\hbar\omega_n\hbar\omega_m]^{1/2}} \left[\int \mathbf{P}^{(nm)}(\mathbf{r}) \cdot \mathbf{D}^{\perp}(\mathbf{r}) \, d\mathbf{r} \right.
$$
$$
\left. - \tfrac{1}{2} \int O_{ij}^{(nm)}(\mathbf{r}, \mathbf{r}') B_i(\mathbf{r}) B_j(\mathbf{r}') \, d\mathbf{r} \, d\mathbf{r}' \right]
$$
$$
+ \sum_m \frac{\omega_n p_m}{[\hbar\omega_n\hbar\omega_m]^{1/2}} \, i \int \mathbf{M}^{[nm]}(\mathbf{r}) \cdot \mathbf{B}(\mathbf{r}) \, d\mathbf{r} \tag{49}
$$

and, unlike equations (13) and (14), there are no inter-atomic contributions. In the electric dipole approximation, equations (48) and (49) simplify to

$$
\dot{q}_n = p_n - \sum_m \frac{p_m}{(\hbar\omega_n\hbar\omega_m)^{1/2}} \boldsymbol{\mu}^{nm} \cdot \mathbf{D}^{\perp}(\mathbf{R}) \tag{50}
$$

$$
\dot{p}_n = -\omega_n^2 q_n + \sum_m \frac{\omega_m\omega_n q_m}{(\hbar\omega_n\hbar\omega_m)^{1/2}} \boldsymbol{\mu}^{nm} \cdot \mathbf{D}^{\perp}(\mathbf{R}) \tag{51}
$$

On the other hand, equations (13) and (14) in the dipole approximation become

$$\dot{q}_n(\alpha) = p_n(\alpha) + \sum_m \frac{q_m(\alpha)\omega_m(\alpha)}{[\hbar\omega_m(\alpha)\hbar\omega_n(\alpha)]^{1/2}} \frac{e}{mc} \mathbf{p}^{mn}(\alpha)\cdot\mathbf{A}(\mathbf{R}_\alpha)$$

$$+ \sum_m \frac{p_m(\alpha)}{[\hbar\omega_m(\alpha)\hbar\omega_n(\alpha]^{1/2}} \frac{e^2}{2mc^2} \delta_{nm}A^2(\mathbf{R}_\alpha)$$

$$+ \tfrac{1}{2} \sum_{m,k,l,\beta} \frac{p_m(\alpha)[p_k(\beta)p_l(\beta) + \omega_k(\beta)\omega_l(\beta)q_k(\beta)q_l(\beta)]}{[\hbar\omega_m(\alpha)\hbar\omega_n(\alpha)\hbar\omega_k(\beta)\hbar\omega_l(\beta)]^{1/2}}$$

$$\times \mu_i^{nm}(\alpha)\mu_j^{kl}(\beta) \frac{[\delta_{ij} - 3(\widehat{\mathbf{R}_\alpha - \mathbf{R}_\beta})_i(\widehat{\mathbf{R}_\alpha - \mathbf{R}_\beta})_j]}{|\mathbf{R}_\alpha - \mathbf{R}_\beta|^3} \tag{52}$$

$$\dot{p}_n(\alpha) = -\omega_n^2(\alpha)q_n(\alpha) + \sum_m \frac{p_m(\alpha)\omega_n(\alpha)}{[\hbar\omega_n(\alpha)\hbar\omega_m(\alpha)]^{1/2}} \frac{e}{mc} \mathbf{p}^{mn}(\alpha)\cdot\mathbf{A}(\mathbf{R}_\alpha)$$

$$- \sum_m \frac{p_m(\alpha)\omega_n(\alpha)}{[\hbar\omega_m(\alpha)\hbar\omega_n(\alpha)]^{1/2}} \frac{e^2}{2mc} \delta_{nm}A^2(\mathbf{R}_\alpha)$$

$$-\tfrac{1}{2} \sum_{m,k,l,\beta} \frac{\omega_n(\alpha)\omega_m(\alpha)q_m(\alpha)[p_k(\beta)p_l(\beta) + \omega_k(\beta)\omega_l(\beta)q_k(\beta)q_l(\beta)]}{[\hbar\omega_m(\alpha)\hbar\omega_n(\alpha)\hbar\omega_k(\beta)\hbar\omega_l(\beta)]^{1/2}}$$

$$\times \mu_i^{nm}(\alpha)\mu_j^{kl}(\beta) \frac{[\delta_{ij} - 3(\widehat{\mathbf{R}_\alpha - \mathbf{R}_\beta})_i(\widehat{\mathbf{R}_\alpha - \mathbf{R}_\beta})_j]}{|\mathbf{R}_\alpha - \mathbf{R}_\beta|^3} \tag{53}$$

As in quantum electrodynamics, the equations of motion that result from the multipolar Hamiltonian (45) are much simpler than those arising from the minimal-coupling Hamiltonian (1); this is especially so for the case of more than one atom. Equations (52) and (53) contain two-center terms, cubic in p's and q's, while (50) and (51) are linear in the coordinates and depend on the transition moments at one center only. Even for a single-center system, the multipolar Hamiltonian (49) has the advantage of being explicitly dependent on the multipole transition moments.

The equations of motion associated with the electromagnetic field are

$$\dot{\mathbf{A}}(\mathbf{r}) = \frac{\partial H_{\text{multi}}}{\partial \mathbf{\Pi}(\mathbf{r})} = 4\pi c^2\mathbf{\Pi}(\mathbf{r}) + \tfrac{1}{2} \sum_{k,l} \frac{p_k p_l + \omega_k\omega_l q_k q_l}{(\hbar\omega_k\hbar\omega_l)^{1/2}} 4\pi c\mathbf{P}^{(kl)}(\mathbf{r}) \tag{54}$$

$$\dot{\mathbf{\Pi}}(\mathbf{r}) = -\frac{\partial H_{\text{multi}}}{\partial \mathbf{A}(\mathbf{r})} = -\frac{1}{4\pi} \text{ curl curl } \mathbf{A}$$

$$-\tfrac{1}{2} \sum_{k,l} \frac{(ip_k q_l\omega_l - ip_l q_k\omega_k + p_k p_l + \omega_k\omega_l q_k q_l)}{(\hbar\omega_k\hbar\omega_l)^{1/2}} \text{ curl } \tilde{\mathbf{M}}^{kl}(\mathbf{r}) \tag{55}$$

where

$$\tilde{M}_i^{kl}(\mathbf{r}) = M_i^{[kl]}(\mathbf{r}) - \int O_{ij}^{(kl)}(\mathbf{r}, \mathbf{r}') B_j(\mathbf{r}') d\mathbf{r}' \tag{56}$$

which is similar to equation (78) of Reference 8. Equation (54) is essentially (31) and relates $\mathbf{D}^\perp$ and $\mathbf{E}^\perp(= -\dot{\mathbf{A}}/c)$. Maxwell's equation as the dynamical equation for the electromagnetic field now arises in the atomic field form with polarization and magnetization currents alone. Equation (55) is

$$(1/c)\dot{\mathbf{D}}^\perp = \text{curl } \mathbf{B} - 4\pi \text{ curl } \tilde{\mathbf{M}} \tag{57}$$

where

$$\begin{aligned}
\tilde{\mathbf{M}}(\mathbf{r}) &= \tfrac{1}{2} \sum_{k,l} \tilde{\mathbf{M}}^{kl}(\mathbf{r}) \frac{(ip_k q_l \omega_l - ip_l q_k \omega_k + p_k p_l + \omega_k \omega_l q_k q_l)}{(\hbar \omega_k \hbar \omega_l)^{1/2}} \\
&= \sum_{k,l} \tilde{\mathbf{M}}^{kl}(\mathbf{r}) a_k^\dagger a_l
\end{aligned} \tag{58}$$

Equation (58) gives the magnetization field in terms of the transition moments, including the diamagnetic term, similar to equation (20) for the electric polarization fields. Finally, introducing the auxiliary field

$$\mathbf{H}(\mathbf{r}) = \mathbf{B}(\mathbf{r}) - 4\pi \tilde{\mathbf{M}}(\mathbf{r}) \tag{59}$$

equation (57) becomes the atomic field equation

$$\text{curl } \mathbf{H} = (1/c)(\partial \mathbf{D}^\perp / \partial t) \tag{60}$$

5. An Application to Interatomic Energies

We use the electric dipole Hamiltonian, equation (46), for two neutral atoms to calculate the interatomic energy shift in the far zone (the Casimir potential[11]) by considering the change in the normal mode frequencies. The Hamiltonian for two atoms (1) and (2) together with the radiation field is

$$\begin{aligned}
H = &H_{\text{rad}} + H_{\text{atoms}} \\
&- \tfrac{1}{2} \sum_{k,l} \frac{p_k(1)p_l(1) + \omega_k(1)\omega_l(1)q_k(1)q_l(1)}{(\hbar \omega_k(1)\hbar \omega_l(1))^{1/2}} \mu^{kl}(1) \cdot \mathbf{D}^\perp(\mathbf{R}_1) \\
&- \tfrac{1}{2} \sum_{k,l} \frac{p_k(2)p_l(2) + \omega_k(2)\omega_l(2)q_k(2)q_l(2)}{(\hbar \omega_k(2)\hbar \omega_l(2))^{1/2}} \mu^{kl}(2) \cdot \mathbf{D}^\perp(\mathbf{R}_2)
\end{aligned} \tag{61}$$

A further canonical transformation within the classical framework, but analogous to that of Craig and Power,[12] can be carried out. It can be defined as in equation (17) with (for one center)

$$
\begin{aligned}
C_{nk} &= \left\{ \exp\left[\frac{i}{\hbar} \frac{\mu \cdot \mathbf{D}^{\perp}(\mathbf{R})}{\Delta} \right] \right\}^{nk} \\
&= \delta_{nk} + \frac{i}{\hbar} \frac{\mu^{nk} \cdot \mathbf{D}^{\perp}(\mathbf{R})}{\omega_n - \omega_k} - \frac{\mu_i^{nl}\mu_j^{lk} D_i(\mathbf{R})D_j(\mathbf{R})}{\hbar^2(\omega_n - \omega_l)(\omega_l - \omega_k)}
\end{aligned}
\tag{62}
$$

Then, ignoring the magnetic terms that arise from the transform of curl $\mathbf{A}$, the leading terms in the transformed Hamiltonian are

$$
\begin{aligned}
H = H_{\mathrm{rad}} &+ H_{\mathrm{atoms}} \\
&- \tfrac{1}{4} \sum_{k,l} \frac{p_k(1)p_l(1) + \omega_k(1)\omega_l(1)q_k(1)q_l(1)}{(\hbar\omega_k(1)\hbar\omega_l(1))^{1/2}} \alpha_{ij}^{(kl)}(1)D_i^{\perp}(\mathbf{R}_1)D_j^{\perp}(\mathbf{R}_1) \\
&- \tfrac{1}{4} \sum_{k,l} \frac{p_k(2)p_l(2) + \omega_k(2)\omega_l(2)q_k(2)q_l(2)}{(\hbar\omega_k(2)\hbar\omega_l(2))^{1/2}} \alpha_{ij}^{(kl)}(2)D_i^{\perp}(\mathbf{R}_2)D_j^{\perp}(\mathbf{R}_2)
\end{aligned}
\tag{63}
$$

In (61) α_{ij} are the anisotropic static polarizabilities of the atoms:

$$
\alpha_{ij}^{(kl)} = \sum_m \mu_i^{km}\mu_j^{ml}\left(\frac{1}{\omega_k - \omega_m} + \frac{1}{\omega_l - \omega_m} \right)
\tag{64}
$$

If the only dynamically significant coordinates for the atoms are p_0, q_0, so that both atoms are in their ground state, we can find the perturbed spectrum of the radiation field due to its weak coupling to these ground state systems. If the radiation field is confined to a big box, its Hamiltonian can be written in terms of the normal mode frequencies $\Omega_\lambda^{(0)}$

$$
H_{\mathrm{rad}} = \sum_\lambda \tfrac{1}{2}(P_\lambda^2 + \Omega_\lambda^{(0)2}Q_\lambda^2)
\tag{65}
$$

where P_λ, Q_λ are the coordinates in the usual transverse vector field decomposition. The perturbed system of radiation can be expressed as

$$
\begin{aligned}
H = \sum \tfrac{1}{2} \{ P_\lambda P_{\lambda'}[\delta_{\lambda\lambda'} &- \tfrac{1}{2}\alpha_{ij}(1)f_{i\lambda}(\mathbf{R}_1)f_{j\lambda}(\mathbf{R}_1) \\
&- \tfrac{1}{2}\alpha_{ij}(2)f_{i\lambda}(\mathbf{R}_2)f_{j\lambda}(\mathbf{R}_2)] + \tfrac{1}{2}\Omega_\lambda^{(0)2}\delta_{\lambda\lambda'}Q_\lambda Q_{\lambda'} \}
\end{aligned}
\tag{66}
$$

The vector functions

$$
\mathbf{f}_\lambda(\mathbf{r}) = \left(\frac{2\pi}{V} \right)^{1/2} i\mathbf{e}^{(\lambda)}(\mathbf{k}_\lambda)e^{i\mathbf{k}_\lambda \cdot \mathbf{r}}
\tag{67}
$$

are the normal mode fields which satisfy

$$
\sum_\lambda \bar{f}_{i\lambda}(\mathbf{r})f_{j\lambda}(\mathbf{r}') = \delta_{ij}^{\perp}(\mathbf{r} - \mathbf{r}')
\tag{68}
$$

It is easy to see that the mode frequencies Ω_λ of this system must satisfy

$$\det[\Omega_\lambda^2 A_{\lambda\lambda'} - \Omega_\lambda^{(0)2}\delta_{\lambda\lambda'}] = 0 \tag{69}$$

where

$$A_{\lambda\lambda'} = \delta_{\lambda\lambda'} - \tfrac{1}{2}\alpha_{ij}(1)\,\check{f}_{i\lambda}(\mathbf{R}_1)f_{j\lambda'}(\mathbf{R}_1) - \tfrac{1}{2}\alpha_{ij}(2)\check{f}_{i\lambda}(\mathbf{R}_2)f_{j\lambda'}(\mathbf{R}_2) \tag{70}$$

For simplicity, we consider the isotropic case from now on; the general result is given in Meath and Power.[13]

The partial trace of the normal mode frequencies, up to a cutoff well below the smallest ω_n above ω_0, can be calculated with the coupling to the atom switched on and switched off. The difference can be calculated using a contour integral method analogous to that of Maradudin, Montroll, and Weiss[14]:

$$\sum(\Omega_\lambda - \Omega_\lambda^{(0)}) = \frac{1}{2\pi}\int_{-\infty}^{\infty} \ln[\Delta(iy)]dy \tag{71}$$

where

$$\Delta(\Omega) = \det[I_{\lambda\lambda'} + R_{\lambda\lambda'}(\Omega)] \tag{72}$$

and the roots of $\Delta(\Omega) = 0$ are the perturbed eigenvalues Ω_λ. $I_{\lambda\lambda'}$ is the unit matrix and

$$R_{\lambda\lambda'}(\Omega) = \frac{2W_{\lambda\lambda'}}{\Omega^{(0)2} - \Omega^2} \tag{73}$$

Comparing equations (69) and (72), we find that

$$W_{\lambda\lambda'} = \tfrac{1}{4}[\alpha(1)\bar{f}_\lambda(\mathbf{R}_1)\cdot f_{\lambda'}(\mathbf{R}_1) + \alpha(2)\,\bar{f}_\lambda(\mathbf{R}_2)\cdot f_{\lambda'}(\mathbf{R}_2)]\Omega^{(0)2} \tag{74}$$

Now equation (71) results in the sum of frequency differences being

$$\frac{1}{2\pi}\int_{-\infty}^{\infty} \text{Tr}\ln[I + R(iy)]dy \approx \frac{1}{2\pi}\int_{-\infty}^{\infty}\text{Tr}(R - \tfrac{1}{2}R^2 + \cdots)dy$$

$$\approx \frac{1}{\pi}\int_{-\infty}^{\infty}\frac{W_{\lambda\lambda'}}{(\Omega_\lambda^{(0)2} + y^2)}dy - \frac{1}{\pi}\int_{-\infty}^{\infty}\frac{W_{\lambda\lambda'}W_{\lambda'\lambda}}{(\Omega_\lambda^{(0)2} + y^2)(\Omega_{\lambda'}^{(0)2} + y^2)}dy \tag{75}$$

As we are concerned with the mutual interaction between atoms 1 and 2, we need consider only the term that depends on $\alpha(1)\alpha(2)$ in the second integral. Using

$$\int_{-\infty}^{\infty}\frac{dy}{(a^2 + y^2)(b^2 + y^2)} = \frac{\pi}{(a + b)ab} \qquad (a, b > 0) \tag{76}$$

we obtain the result

$$- \sum_{\lambda,\lambda'} \frac{W_{\lambda\lambda'} W_{\lambda'\lambda}}{\Omega_\lambda^{(0)} \Omega_{\lambda'}^{(0)} (\Omega_\lambda^{(0)} + \Omega_{\lambda'}^{(0)})}$$

$$= -\tfrac{1}{2} \sum_{\lambda,\lambda'} \frac{\alpha(1)\alpha(2)[\bar{\mathbf{f}}_\lambda(\mathbf{R}_1)\cdot\mathbf{f}_{\lambda'}(\mathbf{R}_1)][\bar{\mathbf{f}}_\lambda(\mathbf{R}_2)\cdot\mathbf{f}_{\lambda'}(\mathbf{R}_2)]\Omega_\lambda^{(0)}\Omega_{\lambda'}^{(0)}}{(\Omega_\lambda^{(0)} + \Omega_{\lambda'}^{(0)})} \tag{77}$$

It is immediately clear from dimensional arguments that this is proportional to $|\mathbf{R}_1 - \mathbf{R}_2|^{-7}$. However, the calculation of the numerical factor is tedious though straightforward.[12] The result is

$$- \frac{23c}{2\pi} \frac{\alpha(1)\alpha(2)}{R^7} \tag{78}$$

Finally, to obtain the energy shift from the neoclassical calculation we simply have to interpret the quantization of wave modes in the standard quantum-electrodynamical way, so that

$$\Delta E = \tfrac{1}{2}\hbar \sum (\Delta\Omega) = - \frac{23\hbar c}{4\pi} \frac{\alpha(1)\alpha(2)}{R^7} \tag{79}$$

which is the Casimir asymptotic potential. One sees that it is possible to hold off the field quantization until the very last step in the calculation.

Appendix

In this appendix we outline a proof of equation (39) of the text. This equation is the transverse part of an identity[10] that equates the nonconvective currents to the sum of the time derivative of the electric polarization and the curl of the magnetization,

$$\mathbf{j}(\mathbf{r}) = -(i/\hbar)[\mathbf{P}(\mathbf{r}), H] + c\ \mathrm{curl}\ \mathbf{M}(\mathbf{r}) \tag{A1}$$

The longitudinal projection of this identity is easily proved. We have

$$\mathbf{P}^{\parallel}(\mathbf{r}) = -\nabla \sum_a \frac{\delta e(a)}{|\mathbf{r} - \mathbf{q}(a)|} \tag{A2}$$

$$\frac{d}{dt} P_i^{\parallel}(\mathbf{r}) = \nabla_i \nabla_j \sum_a \dot{q}_j(a) \frac{e(a)}{|\mathbf{r} - \mathbf{q}(a)|}$$

$$= \sum_a e(a)\dot{q}_j(a)\delta_{ij}^{\parallel}(\mathbf{r} - \mathbf{q}(a)) = j_i^{\parallel}(\mathbf{r}) \tag{A3}$$

To prove the transverse equality, it is most convenient to use the line integral expression[7] for $\mathbf{P}^{\perp}(\mathbf{r})$, which for a single particle centered at $\mathbf{R}$

is

$$P_k^\perp(\mathbf{r}) = -e \sum_a (\mathbf{q}(a) - \mathbf{R})_k \int_0^1 \delta_{jk}^\perp(\mathbf{r} - \mathbf{R} - \lambda(\mathbf{q}(a) - \mathbf{R})) \, d\lambda \quad \text{(A4)}$$

Since

$-(i/\hbar)[P_k^\perp(\mathbf{r}), H]$

$$= -(i/\hbar)[P_k(\mathbf{r}), p^2/2m] + (-i/\hbar)[P_k(\mathbf{r}), (e/mc)\mathbf{p}\cdot\mathbf{A}(\mathbf{q})]$$

$$= -(i/2m\hbar)\{[P_k^\perp(\mathbf{r}), p_i]p_i + p_i[P_k^\perp(\mathbf{r}), p_i]$$

$$+ (2e/c)A_i(\mathbf{q})[P_k^\perp(\mathbf{r}), p_i]\} \quad \text{(A5)}$$

we need to calculate the commutator,

$[P_k^\perp(\mathbf{r}), p_i(a)]$

$$= -e \sum_a \left[[q_j(a) - R_j] \int \delta_{jk}^\perp(\mathbf{r} - \mathbf{R} - \lambda(\mathbf{q}(a) - \mathbf{R}))d\lambda, p_i(a) \right]$$

$$= -i\hbar e \sum_a \frac{\partial}{\partial q_i(a)} (q_j(\alpha) - R_j) \int \delta_{jk}^\perp(\mathbf{r} - \mathbf{R} - \lambda(\mathbf{q}(a) - \mathbf{R}))d\lambda$$

$$= -i\hbar e \sum_a \left\{ \delta_{ij} \int \delta_{jk}^\perp(\mathbf{r} - \mathbf{R} - \lambda(\mathbf{q}(a) - \mathbf{R}))d\lambda \right.$$

$$\left. - (q_j(a) - R_j) \int \lambda \nabla_i \delta_{jk}^\perp(\mathbf{r} - \mathbf{R} - \lambda(\mathbf{q}(a) - \mathbf{R}))d\lambda \right\} \quad \text{(A6)}$$

To simplify the second term in (A6), we write the transverse δ-function as the difference between $\delta_{jk}\delta(\ )$ and $\delta_{jk}^\parallel()$ and then use the cyclic symmetry of $\nabla_i \delta_{jk}^\parallel$. We have

$$-q_j \int \lambda \nabla_i \delta_{jk}^\perp(\mathbf{r} - \mathbf{R} - \lambda(\mathbf{q} - \mathbf{R}))d\lambda$$

$$= - \int \lambda q_k \nabla_i \delta(\mathbf{r} - \mathbf{R} - \lambda(\mathbf{q} - \mathbf{R}))d\lambda$$

$$+ \int \lambda q_j \nabla_j \delta_{ki}^\parallel(\mathbf{r} - \mathbf{R} - \lambda(\mathbf{q} - \mathbf{R}))d\lambda$$

$$= - \int \lambda q_k \nabla_i \delta(\mathbf{r} - \mathbf{R} - \lambda(\mathbf{q} - \mathbf{R}))d\lambda$$

$$- \int \lambda \frac{\partial}{\partial \lambda} \left\{ \delta_{ik}\delta(\mathbf{r} - \mathbf{R} - \lambda(\mathbf{q} - \mathbf{R})) \right.$$

$$\left. - \delta_{ik}^\perp(\mathbf{r} - \mathbf{R} - \lambda(\mathbf{q} - \mathbf{R})) \right\} d\lambda \quad \text{(A7)}$$

The first term in (A6), together with the last term in (A7), gives

$$-i\hbar e \sum_a \int \left(1 + \frac{d}{d\lambda}\right) \delta^{\perp}_{ik}(\mathbf{r} - \mathbf{R} - \lambda(\mathbf{q}(a) - \mathbf{R}))d\lambda$$

$$= -i\hbar e \sum_a \int \frac{d}{d\lambda} \{\lambda \delta^{\perp}_{ik}(\mathbf{r} - \mathbf{R} - \lambda(\mathbf{q}(a) - \mathbf{R}))\}d\lambda$$

$$= -i\hbar e \sum_a \delta^{\perp}_{ik}(\mathbf{r} - \mathbf{q}(a)) \tag{A8}$$

The remaining contribution to (A6) is

$$i\hbar e \sum_a \int \left(q_k \nabla_i + \delta_{ik}\frac{d}{d\lambda}\right) \delta(\mathbf{r} - \mathbf{R} - \lambda(\mathbf{q}(a) - \mathbf{R}))d\lambda \tag{A9}$$

Substituting the sum of (A8) and (A9) in (A5) we have

$$-(i/\hbar)[P^{\perp}_k(\mathbf{r}), H]$$

$$= -(e/2m)\Bigg\{ \sum_a \{[p_i(a)\delta^{\perp}_{ik}(\mathbf{r} - \mathbf{q}(a)) + \delta^{\perp}_{ik}(\mathbf{r} - \mathbf{q}(a))p_i(a)]$$

$$+ (2e/c)A_i(\mathbf{q}(a))\delta^{\perp}_{ik}(\mathbf{r} - \mathbf{q}(a))\}$$

$$- \sum_a \Bigg[\int [p_i(a)(\mathbf{q}(a) - \mathbf{R})_k - p_k(a)(\mathbf{q}(a) - \mathbf{R})_i]$$

$$\times \nabla_i \lambda \delta(\mathbf{r} - \mathbf{R} - \lambda(\mathbf{q}(a) - \mathbf{R}))d\lambda$$

$$+ \int \nabla_i \lambda \delta(\mathbf{r} - \mathbf{R} - \lambda(\mathbf{q}(a) - \mathbf{R}))$$

$$\times [(\mathbf{q}(a) - \mathbf{R})_k p_i(a) - (\mathbf{q}(a) - \mathbf{R})_i p_k(a)]d\lambda$$

$$+ (2e/c)\int [A_i(\mathbf{q})(\mathbf{q}(a) - \mathbf{R})_k - A_k(\mathbf{q})(\mathbf{q}(a) - \mathbf{R})_i]$$

$$\times \nabla_i \lambda \delta(\mathbf{r} - \mathbf{R} - \lambda(\mathbf{q}(a) - \mathbf{R}))d\lambda\Bigg]\Bigg\} \tag{A10}$$

$$= j^{\perp}_k(\mathbf{r}) - c[\text{curl } \mathbf{M}(\mathbf{r})]_k \tag{A11}$$

where

$$\mathbf{j}(\mathbf{r}) = -(e/mc) \sum_a [\dot{\mathbf{q}}(a)\delta(\mathbf{r} - \mathbf{q}(a)) + \delta(\mathbf{r} - \mathbf{q}(a))\dot{\mathbf{q}}(a)] \tag{A12}$$

and

$$\begin{aligned}
\mathbf{M}(\mathbf{r}) = -(e/2c) \sum_a \int \Bigg\{ & \big[(\mathbf{q}(a) - \mathbf{R}) \times \dot{\mathbf{q}}(a) \big] \\
& \times \lambda\delta(\mathbf{r} - \mathbf{R} - \lambda(\mathbf{q}(a) - \mathbf{R})) \\
& + \lambda\delta(\mathbf{r} - \mathbf{R} - \lambda(\mathbf{q}(a) - \mathbf{R})) \\
& \times \big[(\mathbf{q}(a) - \mathbf{R}) \times \dot{\mathbf{q}}(a) \big] \Bigg\} d\lambda
\end{aligned} \tag{A13}$$

It can be shown by taking the curl of equation (A13) that the second term of (A10) is equal to $-c(\text{curl } \mathbf{M})_k$.

References

1. E. T. Jaynes and F. W. Cummings, *Proc. IEEE* **51**, 89 (1963).
2. W. Lamb, *Phys. Rev. A* **134**, 1429 (1964).
3. I. R. Senitzky, *Phys. Rev. A* **6**, 1175 (1972).
4. For a general discussion see *Coherence and Quantum Optics,* Eds. L. Mandel and E. Wolf, Plenum, New York (1973).
5. E. T. Jaynes, reference 4, p. 35.
6. E. A. Power and S. Zienau, *Philos. Trans. R. Soc. London Ser. A* **251**, 427 (1959).
7. R. G. Woolley, *Proc. R. Soc. London Ser. A* **321**, 557 (1971).
8. M. Babiker, E. A. Power, and T. Thirunamachandran, *Proc. R. Soc. London Ser. A* **338**, 235 (1974).
9. E. A. Power and T. Thirunamachandran, *Am. J. Phys.* **46**, 370 (1978).
10. E. A. Power and T. Thirunamachandran, *Mathematika* **18**, 240 (1971).
11. H. B. G. Casimir, *Proc. K. Ned. Akad. Wet.* **60**, 793 (1948).
12. D. P. Craig and E. A. Power, *Int. J. Quant. Chem.* **3**, 903 (1969).
13. W. J. Meath and E. A. Power, *Int. J. Quant. Chem.* **5**, 549 (1971).
14. A. A. Maradudin, E. W. Montroll, and G. W. Weiss, *Solid State Phys. Supp.* **3**, 138 (1963); see also, E. A. Power and T. Thirunamachandran, *Phys. Rev. B* **3**, 3546 (1971).

7

Theory of Natural Line Shape

Luiz Davidovich and
H. M. Nussenzveig

1. Introduction

The quantum-electrodynamical treatment of the emission of light by an atom has been strongly influenced by Weisskopf and Wigner's early contribution[1] to this subject. While their work was highly successful in accounting for the observed line shape, several disturbing theoretical questions concerning the underlying assumptions remained unsettled:

(*i*) An initial state for the system corresponding to an excited atomic eigenstate with no photons present was assumed, which seems quite unphysical. The state preparation and the dependence of the decay on the excitation should be discussed.

(*ii*) It is well known that the exponential decay "ansatz" cannot be valid for all times, although deviations from it are expected to be extremely small for long-lived decaying states such as the atomic ones. However, the range of validity of the exponential decay law should be determined.

(*iii*) The state space was restricted to a two-level atom and to the vacuum and one-photon sectors, without any indication of how to proceed in order to improve the approximation. For such a basic problem as this one, one should start from a clear-cut formulation, and a systematic procedure for deriving corrections to the Weisskopf–Wigner approximation should be given.

Measurements of the Lamb shift in hydrogen provided a sensitive experimental test of the predicted line shape. It was remarked by Lamb[2]

Luiz Davidovich ● Instituto de Física, Pontifícia Universidade Católica do Rio de Janeiro, Brazil *H. M. Nussenzveig* ● Instituto de Física, Universidade de São Paulo, São Paulo, Brazil. Work partially supported by the National Research Council of Brazil

that the Weisskopf–Wigner line shape disagrees with experiment in this case if the usual minimal-coupling interaction Hamiltonian is employed, and that one must use instead the interaction $-e r \cdot \mathbf{E}^\perp$, where $\mathbf{E}^\perp$ is the transverse electric field. As will be seen below, the discrepancy is orders of magnitude larger than the present accuracy in the measurement. Since the minimal-coupling Hamiltonian is widely employed in quantum electrodynamics, this discrepancy should be resolved.

An extensive study of the line-shape problem was made around 1950 by Heitler, Arnous, and collaborators.[3–5] They applied a dressing transformation in order to go over from "bare" states of the system to physical states. They began[3] by imposing unphysical constraints on the dressing transformation; later,[4,5] these constraints were removed, but they employed a cumbersome formalism, and their attention was focused on the evaluation of radiative corrections to the resonant term in a transition between two atomic states. The effect of nonresonant terms was not considered, and there was no discussion of the time development of the system. Similar remarks apply to Low's[6] covariant S-matrix treatment.

The relationship between time evolution and the analytic properties of the S-matrix as a function of energy has been clarified in nonrelativistic potential scattering, where decaying states can be described in terms of propagators associated with complex poles on unphysical sheets.[7] This has also been verified in some models of unstable particles in quantum field theory.

The present paper is a survey of results that were recently obtained[8] in a new treatment of the line-shape problem. We deal with a nonrelativistic hydrogen atom interacting with the quantized radiation field. This is the simplest and best-known atomic system; the corresponding exact transition matrix elements including retardation have been recently determined[9] and they lead to an extremely simple analytic structure of the transition probabilities (Section 2).

We employ Van Hove's[10] resolvent operator approach. An outline of its main features is given in Section 3. In Section 4, we discuss the physical consequences of some of the approximations employed in previous treatments,[11] where various terms in the Hamiltonian are omitted or simplified. We proceed by successive approximations.

We begin with a simplified multilevel model in which "counter-rotating" terms are omitted. This eliminates persistent perturbation effects in the sense of Van Hove (including self-energy and dressing effects) and leads to a decomposition of Hilbert space into sectors, such that the model becomes exactly soluble in the sector of interest. Unlike previously solved models, transitions from the resonant level to other excited states are taken into account.

Employing the exact transition probabilities of Section 2, the resolvent is explicitly obtained and its analytic behavior is found (Section 5). This allows one to describe the time evolution of the system. In particular, we discuss the decay of a Weisskopf–Wigner initial state and the dependence on the excitation in the scattering of a wave packet, thereby determining the range of validity of the exponential decay law within this model (Section 6).

The effect of adding back the counter-rotating terms is considered next. In order to remove the associated persistent effects, we apply a generalized version of a dressing transformation proposed by Faddeev,[12] which is defined order by order in perturbation theory (Section 7). The previously described resolvent operator method can then be applied to the transformed Hamiltonian, which generates the dynamics of "dressed" states.

The results obtained by applying this procedure to second order in the coupling constant, in dipole approximation, are described in Section 8. The dressing transformation yields the nonrelativistic Lamb-shift correction to the ground-state energy. The corresponding Lamb-shift corrections to excited-state energies appear, as they should, in the poles of the resolvent on unphysical sheets.

Finally, in Section 9, we discuss the relation between the minimal coupling and $-e\mathbf{r}\cdot\mathbf{E}^{\perp}$ interaction Hamiltonians, as well as the results they yield for the line shape. While these results are equivalent when both the resonant and all nonresonant terms are included, this is not so when, as is usually done, only the resonant term is taken into account. The background contribution due to transitions to all nonresonant levels can be explicitly computed with the help of the Coulomb Green's function. For the Lamb-shift transition, the background correction to the minimal-coupling resonant term is important, and it resolves the discrepancy observed by Lamb. For the Lyman-α line, although the correction is much smaller, it is the minimal-coupling resonant term that yields better results. The evaluation of corrections to the line shape should become of increasing importance as the accuracy of Lamb-shift measurements increases.

2. The Model

We consider a nonrelativistic hydrogen atom (recoil is neglected) interacting with the quantized radiation field. The Hamiltonian of the system is (we take $\hbar = c = 1$ throughout)

$$H = H_A + H_F + H_{I,1} + H_{I,2} \qquad (2.1)$$

with

$$H_A = \mathbf{p}^2/2m - e^2/r \tag{2.2}$$

$$H_F = (1/8\pi) \int d^3r \, [(\mathbf{E}^\perp)^2 + (\boldsymbol{\nabla} \times \mathbf{A})^2] \tag{2.3}$$

where $\mathbf{A}$ is the vector potential in the Coulomb gauge and $\mathbf{E}^\perp$ is the transverse electric field; the linear and quadratic parts of the interaction Hamiltonian are respectively given by

$$H_{I,1} = -(e/m)\mathbf{p}\cdot\mathbf{A}(\mathbf{r}) \tag{2.4}$$

$$H_{I,2} = (e^2/2m)\mathbf{A}^2(\mathbf{r}) \tag{2.5}$$

We rewrite the atomic Hamiltonian in terms of the hydrogen atom stationary states $|n\rangle$ (where n stands for a complete set of quantum numbers) and the corresponding energies E_{0n} as

$$H_A = \sum_n E_{0n} |n\rangle\langle n| \tag{2.6}$$

where the summation is to be understood as integration for eigenstates in the continuum. We will be concerned, however, mainly with the discrete spectrum.

We employ the multipole expansion[13]

$$\mathbf{A}(\mathbf{r}) = 2 \sum_{\tau=0}^{1} \sum_{J=1}^{\infty} \sum_{M=-J}^{J} \int_0^\infty dk \, k^{1/2} a_{JM\tau}(k)\mathbf{A}_{JM\tau}(k, \mathbf{r}) + \text{H. c.} \tag{2.7}$$

in terms of the usual basis,[13] where $\mathbf{A}_{JM\tau}$ represents an electric ($\tau = 0$) or magnetic ($\tau = 1$) multipole field of order 2^J. The operator $a_{JM\tau}(k)$ annihilates photons of frequency k with the set of quantum numbers

$$\beta = (JM\tau) \tag{2.8}$$

so that

$$[a_\beta(k), a_{\beta'}{}^\dagger(k')] = \delta_{\beta\beta'}\delta(k - k') \tag{2.9}$$

The field Hamiltonian becomes (after zero-point energy subtraction)

$$H_F = \sum_\beta \int_0^\infty dk \, k a_\beta{}^\dagger(k)a_\beta(k) \tag{2.10}$$

The exact matrix elements $\langle n|\mathbf{p}\cdot\mathbf{A}_\beta|m\rangle$ for the nonrelativistic hydrogen atom have been evaluated by Moses.[9] To express $H_{I,1}$ in terms of them, it is convenient to adopt "atomic units" for the frequency k, measuring it in units of the inverse Bohr radius a_B^{-1}, i.e., setting

$$a_B^{-1} \equiv \alpha m = 1 \tag{2.11}$$

where α is the fine-structure constant. We then have

$$H_{I,1} = \lambda^{1/2} \sum_{n,m,\beta} \int_0^\infty dk \, k^{1/2} f_{nm\beta}(k) a_\beta(k) |n\rangle\langle m| + \text{H. c.} \qquad (2.12)$$

where we have defined a coupling constant [in the units (2.11)]

$$\lambda = (e/m)^2 = \alpha^3 \quad (\approx 4 \times 10^{-7}) \qquad (2.13)$$

Let $n = (N_n, j_n, M_n)$, where N_n, j_n, and M_n are the principal quantum number, angular momentum, and magnetic quantum number of the state n, respectively. The matrix elements $f_{nm\beta}$ vanish unless they fulfill the following exact selection rules, which follow from angular momentum and parity conservation:

$$|j_n - j_m| \leq J \leq j_n + j_m \qquad M = M_n - M_m$$

$$J + j_n + j_m \equiv \tau \pmod 2 \qquad (2.14)$$

Under these conditions, it can be shown[8] that $|f_{nm\beta}(k)|^2$ (derived from the results of Reference 9) is a rational function of k^2 and has the following remarkably simple form:

$$|f_{nm\beta}(k)|^2 = P_{nm\beta}(k)(k^2 + K_{nm}^2)^{-p_{nm}} \qquad (2.15)$$

where $P_{nm\beta}(k)$ is a polynomial (in k^2) with real coefficients, that are $O(1)$ in the units (2.11), p_{nm} is an integer ≥ 4, such that

$$|f_{nm\beta}(k)|^2 = O(k^{-8}) \qquad \text{as } k \to \infty \qquad (2.16)$$

and $K_{nm}^{-1} = O(1)$ is an atomic transition radius of the order of the Bohr radius. For example, for the Lyman-α and Lyman-β lines (electric dipole transitions), $|f_{nm\beta}(k)|^2$ is proportional to, respectively,

$$[k^2 + (\tfrac{3}{2})^2]^{-4}, \qquad [(\tfrac{4}{3})^2 + 2k^2]^2[k^2 + (\tfrac{4}{3})^2]^{-6}$$

The functions $|f_{nm\beta}(k)|^2$ play the role of exact atomic form factors, introducing a natural cutoff at distances of the order of the Bohr radius. The Yukawa-like denominators in (2.15) reflect the exponential falloff of the bound-state wave functions. The extremely simple analytic properties of (2.15) play an important role in the soluble model discussed below.

3. The Resolvent Operator

Our treatment is based on the resolvent operator method, as developed by Van Hove.[10] We briefly recall its main features.

The resolvent operator $\mathcal{G}(z)$ associated with the Hamiltonian $H = H_0 + H_I$ is an operator-valued function of the complex variable z defined, for Im $z \neq 0$, by

$$\mathcal{G}(z) = (z - H)^{-1} \qquad (3.1)$$

It is expected to be holomorphic for Im $z \neq 0$. The time evolution operator may be expressed in terms of $\mathcal{G}(z)$ as

$$\exp(-iHt) = -(1/2\pi i) \int_C \exp(-izt)\mathcal{G}(z)dz \qquad (t > 0) \qquad (3.2)$$

where C is a straight line taken above the real axis from $-\infty + i\epsilon$ to $\infty + i\epsilon$ ($\epsilon \geq 0$).

The operator $\mathcal{G}(z)$ can be split into a diagonal part with respect to H_0, $\mathcal{D}(z)$, and a nondiagonal part with respect to H_0, $\mathcal{N}(z)$:

$$\mathcal{G}(z) = \mathcal{D}(z) + \mathcal{N}(z) \qquad (3.3)$$

where, for every eigenstate $|\alpha\rangle$ of H_0,

$$H_0|\alpha\rangle = E_{0\alpha}|\alpha\rangle \qquad (3.4)$$

we have

$$\mathcal{D}(z)|\alpha\rangle = \mathcal{D}_\alpha(z)|\alpha\rangle \qquad (3.5)$$

The eigenvalue $\mathcal{D}_\alpha(z)$ determines the persistence amplitude of the state $|\alpha\rangle$, i.e., if we start out from the initial state $|\alpha\rangle$ at $t = 0$, the probability amplitude to find the system in the state $|\alpha\rangle$ at time t is given by

$$\langle\alpha|\exp(-iHt)|\alpha\rangle = -(1/2\pi i) \int_C \exp(-izt)\mathcal{D}_\alpha(z)dz \qquad (3.6)$$

This amplitude therefore depends crucially on the analytic properties of $\mathcal{D}_\alpha(z)$.

The unperturbed resolvent operator $\mathcal{G}_0(z)$ is defined by

$$\mathcal{G}_0(z) = (z - H_0)^{-1} \qquad (3.7)$$

The analogue $\Sigma(z)$ of Dyson's mass operator[14] satisfies

$$\mathcal{D}(z) = \mathcal{G}_0(z) + \mathcal{D}(z)\Sigma(z)\mathcal{G}_0(z) \qquad (3.8)$$

and it can be pictured diagrammatically by

$$\Sigma(z) = (H_I + H_I\mathcal{D}(z)H_I + \cdots)_{\text{i.d.}} \qquad (3.9)$$

where the index "i.d." stands for a summation over all irreducible diagonal[10] diagrams.

As z approaches a point E on the real axis, we have[10]

$$\lim_{z \to E \pm i0} \Sigma(z) = \Delta(E) \mp \tfrac{1}{2}i\Gamma(E) \qquad (3.10)$$

where $\Gamma(E)$ is a positive semidefinite operator, $\Gamma(E) \geq 0$. Thus, denoting by an index α the eigenvalues of diagonal operators in the state $|\alpha\rangle$

$$\lim_{z \to E \pm i0} \mathscr{D}_\alpha(z) = [E - E_{0\alpha} - \Delta_\alpha(E) \pm \tfrac{1}{2}i\Gamma_\alpha(E)]^{-1} \qquad (3.11)$$

If $\Gamma_\alpha(E) \neq 0$, we see that E lies on a cut of $\mathscr{D}_\alpha(z)$, which is holomorphic for Im $z \neq 0$ (physical sheet).

The Weisskopf–Wigner approximation would correspond to

$$\mathscr{D}_\alpha(z) = (z - E_\alpha + \tfrac{1}{2}i\Gamma_\alpha)^{-1} \qquad (3.12)$$

which would violate the analyticity on the physical sheet. By comparison with (3.11), we see that $\Delta(E)$ and $\Gamma(E)$ in (3.10) play the roles of level shift and level width operators, respectively.

The nondiagonal part $\mathscr{N}(z)$ in (3.3) may be written as

$$\mathscr{N}(z) = \mathscr{D}(z)\mathscr{U}(z)\mathscr{D}(z) \qquad (3.13)$$

where the transition operator $\mathscr{U}(z)$ is diagrammatically represented by

$$\mathscr{U}(z) = [H_I + H_I\mathscr{D}(z)H_I + \cdots]_{\text{i.n.d}} \qquad (3.14)$$

summed over all irreducible nondiagonal diagrams (i.n.d.).[10] As will be seen below, $\mathscr{U}(z)$ determines the transition amplitude in scattering processes.

Let $|\alpha\rangle$ stand, as in (3.4), for an eigenstate of H_0 with unperturbed energy $E_{0\alpha}$, and let us assume that the equation [cf. (3.11)]

$$E - E_{0\alpha} - \Delta_\alpha(E) = 0 \qquad (3.15)$$

has one and only one real root $E = E_\alpha$. Then, there are only three possibilities:

$$(i) \quad \Gamma_\alpha(E_\alpha) \neq 0 \qquad (3.16)$$

In this case, by (3.11), $E = E_\alpha$ lies on a cut of $\mathscr{D}_\alpha(z)$. Van Hove[10] calls this a *dissipative state*; typically, it decays in the presence of the interaction, and $\Delta_\alpha(E_\alpha)$ and $\Gamma_\alpha(E_\alpha)$ represent the energy shift and linewidth due to the interaction. This situation is well illustrated by a Weisskopf–Wigner initial state,

$$|\alpha\rangle = |r; 0\rangle = |\mathrm{r}\rangle|0) \qquad (3.17)$$

where $|r\rangle$ is an excited state of H_A and $|0)$ denotes the photon vacuum (we employ round brackets for photon state vectors). This will be discussed below in Section 6.1.

$$(ii) \quad \Gamma_\alpha(E) = 0 \quad \text{for all real } E \qquad (3.18)$$

In this case, $\mathscr{D}_\alpha(z)$ has a simple pole at $E = E_\alpha$ and no cut. The state $|\alpha\rangle$ is classified by Van Hove as *asymptotically stationary*,[10] because its asymptotic evolution is not affected by the interaction. The interaction produces only transient effects, as in ordinary scattering from a short-range potential. An example will be provided by the scattering of a one-photon wave packet from the ground state for the exactly soluble model discussed in Section 6.2.

$$(iii) \quad \Gamma_\alpha(E_\alpha) = 0, \qquad \text{but } \Gamma_\alpha(E) \neq 0 \text{ for some real } E \qquad (3.19)$$

In this case, the state $|\alpha\rangle$ is not asymptotically stationary. The interaction produces *persistent perturbation effects*; besides self-energy effects, they typically include "dressing" ("cloud") effects. Thus (cf. Section 8), the state $|\alpha\rangle = |1; 0\rangle = |1\rangle|0\rangle$, where $|1\rangle$ is the ground state of H_A, is not asymptotically stationary; the interacting ground state of the system is "dressed."

If $|\alpha\rangle$ and $|\alpha'\rangle$ are asymptotically stationary, the S-matrix element $\langle\alpha|S|\alpha'\rangle$ between these states can be shown to exist (with no need to adopt the unphysical procedure of adiabatic switching of the interaction). It is then given by

$$\langle\alpha|S|\alpha'\rangle = \delta(\alpha - \alpha') - 2i\pi\delta(E_\alpha - E_{\alpha'})\mathcal{U}_{\alpha\alpha'}(E_\alpha) \qquad (3.20)$$

where

$$\mathcal{U}_{\alpha\alpha'} = \langle\alpha|\mathcal{U}|\alpha'\rangle \qquad (3.21)$$

is the transition amplitude matrix with $\mathcal{U}$ given by (3.13) and (3.14). The corresponding differential cross section for one-photon incidence is given by

$$\sigma_{\alpha\alpha'} = (2\pi)^4|\mathcal{U}_{\alpha\alpha'}(E_\alpha)|^2\rho_\alpha(E_\alpha) \qquad (3.22)$$

where ρ_α is the density of final states.

4. Possible Approximations

Our approach is based on the idea of successive approximations, retaining first only a part of the Hamiltonian, such that the problem becomes exactly soluble, and then adding back the omitted pieces of the Hamiltonian and investigating how they affect the solution. A somewhat related approach for Weisskopf–Wigner-type theories has been advocated by Grimm and Ernst.[15] We start with a physical discussion of some of the main approximations that have previously been employed,[11] as well as an extended version to be employed in Section 5.

(I) Finite Number of Atomic States. This amounts to cutting off the spectrum of H_A beyond some discrete level N, so that, in (2.6) and (2.12),

$$(n, m) \leq N \qquad (4.1)$$

In particular, if only two states are kept, this is the two-level atom. The accumulation point at the ionization threshold as well as transitions to the continuum are eliminated by this approximation.

(II) Generalized RWA ("Rotating-Wave Approximation"). We consider a specific excited state $|r\rangle$, to be kept fixed, where "r" stands for "resonant," because this state is supposed to be selectively singled out by resonant processes, as will be seen below. Keeping only those terms in (2.12) where one of the levels involved in the transition is r, the generalized RWA corresponds to the following choice of interaction Hamiltonian:

$$H_I = \lambda^{1/2} \sum_{n\beta} \int_0^\infty dk\, k^{1/2} f_{rn\beta}(k) a_\beta(k) |r\rangle\langle n| + \text{H. c.} \qquad (4.2)$$

where the counter-rotating terms, which differ from those in (4.2) by the interchange of r and n within the summation, are neglected.

This is an extended version of the usual RWA, to which it reduces for a two-level atom. It corresponds to coupling the *absorption* of a photon only with transitions that *end* in r, and *emission* of a photon only with those that *start* from r, whether the transitions are to levels above or below r.

If we call "resonant atomic excitation" the occupation of level r (so that the atom is not "resonantly excited" whenever it occupies a level $n \neq r$), the generalized RWA leads to an additional conservation law, corresponding to the conservation of the number of photons plus the resonant atomic excitation. This is represented by the operator

$$\mathscr{E} = \tfrac{1}{2}|r\rangle\langle r| - \tfrac{1}{2}\sum_{n\neq r}|n\rangle\langle n| + \sum_\beta \int_0^\infty dk\, a_\beta^\dagger(k) a_\beta(k) \qquad (4.3)$$

which commutes with

$$H = H_0 + H_I = H_A + H_F + H_I \qquad (4.4)$$

i.e.,

$$[\mathscr{E}, H] = 0 \qquad (4.5)$$

This leads to a splitting of Hilbert space into sectors, allowing an exact solution in the sector of interest, as will be seen below.

Furthermore, (4.2) allows a transition from the ground state ($n = 1$) to level r only in the presence of a photon, so that the unperturbed ground state of the system

$$|1; 0\rangle = |1\rangle |0\rangle \qquad (4.6)$$

is also the interacting ground state, an eigenstate of H in the sector $\mathscr{E} =$

$-\frac{1}{2}$ [so long as (5.6) below is satisfied]. Thus, the generalized RWA excludes persistent perturbation effects.

(III) Dipole Approximation. This amounts to substituting, in (2.4) and (2.5),

$$\mathbf{A}(\mathbf{r}) = \mathbf{A}(0) \qquad (4.7)$$

As a consequence of this, only electric dipole waves remain coupled to the atom, i.e., the photon index ranges only over the values

$$J = 1; \qquad M = 0, \pm 1; \qquad \tau = 0 \qquad (4.8)$$

At the same time, this corresponds to neglecting retardation over atomic dimensions, so that the electric dipole form factors in (2.15) go over into their $k \to 0$ limit, which is just a constant. This would lead to divergent integrals, so that this approximation is often coupled with (IV).

(IV) Sharp Cutoff. To avoid divergences in the dipole approximation, one replaces the effect of the atomic form factors by a sharp cutoff,

$$\rho(k) = \theta(K - k) \qquad (4.9)$$

where θ is the Heaviside step function and the cutoff parameter K is of the order of the inverse Bohr radius, $K \sim a_B^{-1}$.

The combination of approximations (III) and (IV), although it has often been employed in conjunction with two-level atomic models, leads to spurious effects. It introduces a spurious pole of the resolvent on the real axis, which gives rise to a nondecaying, nonergodic contribution to the time evolution[8] for the state (3.17). This does not happen for a smooth form factor such as (2.15).

We do not employ either (III) or (IV) in the soluble model that will now be discussed.

5. Exactly Soluble Multilevel Model

This model is defined by the Hamiltonian (4.4),

$$H = \sum_n E_{0n} |n\rangle\langle n| + \sum_\beta \int_0^\infty dk\, k a_\beta^\dagger(k) a_\beta(k)$$
$$+ \lambda^{1/2} \sum_{n\beta} \int_0^\infty dk\, k^{1/2} [f_{rn\beta}(k) a_\beta(k) |r\rangle\langle n| + \text{H. c.}] \qquad (5.1)$$

where $n \le N$, and β also ranges only over a finite set of values, due to the selection rules (2.14). The coefficients $f_{rn\beta}(k)$ are taken to be the exact hydrogen atom matrix elements. This corresponds to making approximations (I) and (II) of Section 4 and neglecting the quadratic interaction Hamiltonian (2.5).

The transitions allowed by the interaction Hamiltonian in (5.1) are schematically represented in Figure 1. In contrast with the multilevel model treated by Davies,[16] the present model allows for transitions (within the generalized RWA) between the resonant level and other excited levels, but it does not take into account the decay of the other levels.

We restrict our discussion to the Hilbert space sector associated with the eigenvalue $\mathscr{E} = \frac{1}{2}$ of the operator (4.3). Any state vector in this sector is of the form

$$|\psi\rangle = u|r; 0\rangle + |\Phi\rangle \tag{5.2}$$

where

$$|\Phi\rangle = \sum_{n\neq r, \beta} \int_0^\infty dk\, \varphi_{n\beta}(k)|n; 1\beta, k\rangle \tag{5.3}$$

$$|r; 0\rangle = |r\rangle|0\rangle, \qquad |n; 1\beta, k\rangle = a_\beta^\dagger(k)|n\rangle|0\rangle \tag{5.4}$$

$$\langle\psi|\psi\rangle = |\mu|^2 + \sum_{n\neq r, \beta} \int_0^\infty |\varphi_{n\beta}(k)|^2\, dk = 1 \tag{5.5}$$

The state vector $|\psi\rangle$ evolves in the 0 photon + 1 photon subspace. Thus, cascade and multiphoton processes are excluded, but the effects of transitions to other levels on the excitation and decay of the resonant level are partially taken into account.

The condition

$$\frac{\lambda}{E_{0r}} \sum_{n\neq r, \beta} \int_0^\infty \frac{|f_{rn\beta}(k)|^2}{k + E_{0n}} k\, dk < 1 \tag{5.6}$$

which follows for this model from the smallness of the coupling constant (2.13), ensures that H has only a continuous spectrum,[17] ranging from the unperturbed ground-state energy E_{01} to infinity. We take the unper-

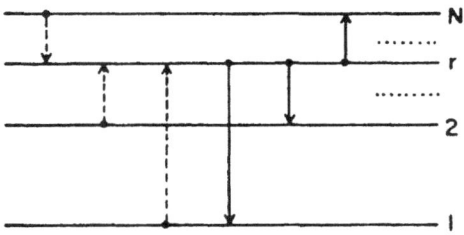

FIGURE 1. Transitions taken into account by the soluble *N*-level model: r is the resonant level; $--\to$ corresponds to absorption, $\longrightarrow$ to emission.

turbed ground-state energy as the zero level of energy,

$$E_{01} = 0 \qquad (5.7)$$

The resolvent then has a cut along the real axis ranging from 0 to ∞.

The matrix elements of the resolvent in the sector under consideration follow from the identity

$$\mathscr{G}(z) = \mathscr{G}_0(z) + \mathscr{G}(z)H_I\mathscr{G}_0(z) \qquad (5.8)$$

They are explicitly given by[8]

$$\langle r; 0|\mathscr{G}(z)|r; 0\rangle = \langle r; 0|\mathscr{D}(z)|r; 0\rangle = \mathscr{D}_{r;0}(z) = 1/D(z) \qquad (5.9)$$

$$\langle r; 0|\mathscr{G}(z)|\Phi\rangle = \frac{\lambda^{1/2}}{D(z)} \sum_{n\neq r,\beta} \int_0^\infty \frac{k^{1/2} f_{rn\beta}^*(k)\varphi_{n\beta}(k)}{z - k - E_{0n}} dk \qquad (5.10)$$

$$\langle \Phi|\mathscr{G}(z)|\Phi'\rangle = \sum_{n\neq r,\beta} \int_0^\infty \frac{\varphi_{n\beta}^*(k)\varphi_{n\beta}'(k)}{z - k - E_{0n}} dk$$

$$+ \frac{\lambda}{D(z)} \sum_{n\neq r,\beta} \int_0^\infty \frac{k^{1/2}\varphi_{n\beta}^*(k) f_{rn\beta}(k)}{z - k - E_{0n}} dk$$

$$\times \sum_{m\neq n,\beta} \int_0^\infty \frac{k^{1/2} f_{rm\beta}^*(k)\varphi_{m\beta}'(k)}{z - k - E_{0m}} dk \qquad (5.11)$$

where

$$D(z) = z - E_{0r} - \frac{1}{2\pi} \sum_{n\neq r,\beta} \int_0^\infty \frac{\Gamma_{rn\beta}(k)}{z - k - E_{0n}} dk \qquad (5.12)$$

$$\Gamma_{rn\beta}(k) = 2\pi\lambda k |f_{rn\beta}(k)|^2 \qquad (5.13)$$

The analytic properties of $\mathscr{D}_{r;0}(z)$ and of the other matrix elements of the resolvent are therefore determined by those of the denominator function $D(z)$. This function can be explicitly computed[8] from (2.15). It is given by

$$D(z) = z - E_{0r} - \frac{1}{2\pi} \sum_{n\neq r,\beta} \Gamma_{rn\beta}(z - E_{0n}) \ln[(E_{0n} - z)/K_{rn}]$$

$$+ \lambda \sum_{n\neq r,\beta} \frac{Q_{rn\beta}(z - E_{0n})}{[(z - E_{0n})^2 + K_{rn}^2]^{p_{rn}}} \qquad (5.14)$$

where, on the physical sheet,

$$\ln\left(\frac{E_{0n} - E \mp i0}{K_{rn}}\right) = \ln\left|\frac{E_{0n} - E}{K_{rn}}\right| \mp i\pi\theta(E - E_{0n}) \qquad (5.15)$$

θ being the Heaviside step function, and $Q_{rn\beta}(z)$ is a polynomial with real coefficients that are $O(1)$ in the units (2.11), with degree $Q_{rn\beta} \leq 2p_{rn} - 1$. Thus,

$$D(z) = O(z) \qquad (|z| \to \infty) \tag{5.16}$$

The function $D(z)$ is holomorphic and zero-free on the physical sheet. The coefficients of the polynomial $Q_{rn\beta}(z)$ follow from (2.15), (5.13), (5.14), and the condition that $D(z)$ has no singularities at $z = E_{0n} \pm iK_{rn}$ on the physical sheet.

According to (5.14), $D(z)$ has a logarithmic branch point at each unperturbed bound-state energy E_{0n}, except the resonant one, $n = r$. The logarithmic terms, while reminiscent of the "Bethe-log" contributions to the Lamb shift (cf. Section 8), are exact within this model.

6. Applications to Decay and Resonance Scattering

We now apply the exactly soluble model of Section 5 to investigate the decay of a Weisskopf–Wigner initial state and the resonance scattering of a wave packet.

6.1. Decay of a Weisskopf–Wigner Initial State

The initial state $|r; 0\rangle$ defined by (5.4) is a dissipative state in the sense of Van Hove [cf. (3.16) and (5.15)]. According to (3.6) and (5.9),

$$\langle r; 0 | \exp(-iHt) | r; 0 \rangle = -\frac{1}{2\pi i} \int_C \frac{\exp(-izt)}{D(z)} \, dz \tag{6.1}$$

Let us now deform the path C in the manner indicated in Figure 2, into a series of paths $C_1, C_2, \ldots, C_N$ directed parallel to the bisector of the fourth quadrant, such that the path C_j winds around the branch point E_{0j}. Thus, the left-hand side of C_j and the right-hand side of C_j are on different Riemann sheets, but the left-hand side of each path is on the same sheet as the right-hand side of the preceding one. These deformations are allowed by (5.16), with no contribution from portions of the "circle at infinity."

Each time the path winds around a branch point E_{0j} in the indicated manner, a new term

$$+ i \sum_\beta \Gamma_{rj\beta}(z - E_{0j})$$

is added to the determination of $D(z)$ in the corresponding Riemann sheet [cf. (5.14) and (5.15)]. On each sheet, the integrand has a finite number of poles, the positions of which can be determined[8] from (5.14). With the above choice of contour it may be shown[8] that only one pole z_r

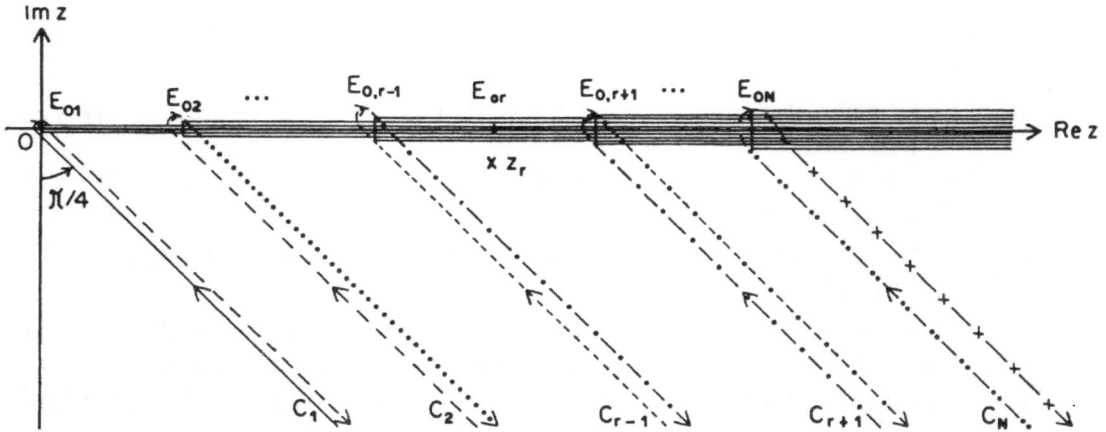

FIGURE 2. Equivalent path of integration for (6.1). The positive real axis is covered by a series of superimposed branch cuts associated with the threshold branch points, and the path winds around successive branch points; portions belonging to the same Riemann sheet are similarly represented. The pole z_r of (6.2) is indicated.

(indicated in Figure 2) will be crossed, with

$$z_r = E_r - \tfrac{1}{2}i\Gamma_r = E_{0r} + \frac{1}{2\pi} \sum_{n \neq r, \beta} P \int_0^\infty \frac{\Gamma_{rn\beta}(k)}{E_{0r} - k - E_{0n}} \, dk$$

$$- \tfrac{1}{2}i \sum_{\beta, n=1}^{r-1} \Gamma_{rn\beta}(E_{0r} - E_{0n}) + O(\lambda) \qquad (6.2)$$

where only those thresholds crossed up to E_{0r} contribute to the imaginary part, and P denotes Cauchy's principal value.

For times much longer than the optical period ($E_{0r} t \gg 1$), the asymptotic behavior of (6.1) is given by[8]

$$\langle r; 0 | \exp(-iHt) | r; 0 \rangle \approx [1 + O(\lambda^2)] \exp(-iE_r t - \tfrac{1}{2}\Gamma_r t)$$

$$+ \sum_{n \neq r, \beta} A_{rn\beta} \, t^{-2\gamma_{rn\beta}} \exp(-iE_{0n}t) \qquad (6.3)$$

where $A_{rn\beta}$ are constants and

$$\gamma_{rn\beta} = J_{rn\beta} + \tau_{rn\beta} \qquad (6.4)$$

where $J_{rn\beta}$ and $\tau_{rn\beta}$ are associated with the multipolarity of the transition [cf. (2.8)]; for electric dipole transitions, $\gamma_{rn\beta} = 1$.

The first term in (6.3) corresponds to the Wigner–Weisskopf exponential "ansatz," with the level shift $E_r - E_{0r}$ given by the principal-value integral in (6.2) (which already contains, as we have seen, pieces of "Bethe-log"-like contributions). The half-width Γ_r, according to (6.2) is the sum of the partial widths for transitions to all *lower* levels, as it should be. Both the level shift and half-width include contributions that were not taken into account in previous exactly solved models (in particular, those involving two-level atoms).

The terms in the second line of (6.3) represent corrections to exponential decay, arising from the integrals around the cuts C_j in Figure 2. A rigorous estimate[8] shows that, for electric dipole transitions, such corrections are negligible up to times t_0 such that

$$t_0^2 \exp(-\tfrac{1}{2}\Gamma_r t_0) \sim (2e^2/3\pi)\langle r|r^2|r\rangle \qquad (6.5)$$

For the Lyman-α line, this yields a value of t_0 of the order of 96 lifetimes, so that corrections to the Weisskopf–Wigner "ansatz" in this case are extremely small. It is interesting to note that such corrections behave as t^{-2} for $t \to \infty$, just like a free-photon wave packet when account is taken of causal propagation.[18] Analogous features have been found for decay involving massive particles,[7] suggesting that limitations in the validity of the exponential decay law are to be traced to the limitations in the validity of the particle concept itself when applied to unstable particles. In order to discuss the observability of such deviations, however, one would need a theoretical analysis of the measurement process.

6.2. Resonance Scattering of a Wave Packet

The dependence of the decay on the excitation in the present model can be investigated by discussing the scattering of a photon wave packet by the atom. In order for the model to remain reasonably realistic, we must assume that the mean energy $\bar{E}$ of the incident wave packet is close to resonance [cf. (6.7) below], so that we are discussing resonance fluorescence.

As a consequence of the generalized RWA employed in our model, the state $|1; 1\beta, k\rangle$ defined by (5.4) (atom in the ground state + 1 photon) is asymptotically stationary in the sense of Van Hove, so that the incident wave takes the form

$$|\Phi\rangle = \int_0^\infty dk\, \varphi(k)|1; 1\beta, k\rangle \qquad (6.6)$$

where $\varphi(k)$ is associated with a mean energy $\bar{E}$ and a half-width γ such that

$$|\bar{E} - E_{0r}| \ll |E_{0,r} - E_{0,r\pm1}|, \qquad \gamma \ll |E_{0,r} - E_{0,r\pm1}| \qquad (6.7)$$

The excitation amplitude of the atom is given by [cf. (5.10)]

$$\langle r; 0|\exp(-iHt)|\Phi\rangle = -(1/2\pi i)\int_C \exp(-izt)\langle r; 0|\mathcal{G}(z)|\Phi\rangle\, dz \qquad (6.8)$$

By suitable choice of $\varphi(k)$ (e.g., a Lorentzian wave packet), this can be treated[8] similarly to (6.1). Typically, the excitation probability is characterized by a rise time T_{rise} and by a decay time T_{decay} such that[5,7]

$$T_{\text{rise}} \sim \min(\gamma^{-1}, \Gamma_r^{-1}); \qquad T_{\text{decay}} \sim \max(\gamma,^{-1}, \Gamma_r^{-1}) \qquad (6.9)$$

For excitation by a broad line ($\gamma \gg \Gamma_r$), the excitation and the decay may be regarded as to some extent independent processes; for a Lorentzian wave packet, the Weisskopf–Wigner state is excited with a probability $\sim(\Gamma_r/\gamma)^2$ and it decays with the natural line shape.

The scattered wave packet, as well as the emitted line shape, can also be obtained[8] with the help of (5.11). The asymptotic behavior for large times is found to depend very strongly on the excitation.

7. Persistent Effects and Dressing Transformation

When the counter-rotating terms that were omitted from the Hamiltonian to yield the exactly soluble model are reintroduced, drastic changes take place: these terms lead to virtual transitions from the unperturbed ground state $|1; 0\rangle$ to excited states with the emission of a photon, followed by the reverse transition in which the photon is absorbed. This produces persistent perturbation effects in the sense of Van Hove, i.e., both self-energy and cloud effects.

These effects are manifested in several ways: $|1; 0\rangle$ is no longer the ground state of the interacting system (it gets "dressed by a photon cloud"), and the states $|1; 1\beta, k\rangle$ are no longer asymptotically stationary, so that they cannot be employed as a basis to build up wave packets in the treatment of resonance scattering.

In order to deal with this situation, we first apply a dressing transformation in order to get rid of the persistent effects. After this is done, the resolvent operator method can in principle be employed as before, with the transformed Hamiltonian now generating the dynamics of "dressed" states of the system. However, exact results are no longer obtained: the transformation is defined order by order in perturbation theory; we discuss the results obtained to second order. Mass renormalization must also be performed.

Attempts to perform such a transformation were made by Heitler, Arnous, and collaborators in a series of papers,[3] but they met with difficulties, because they attempted to solve the unphysical problem of dressing also excited atomic states. In their subsequent work,[4,5] the transformation was suitably restricted, but no explicit expressions for it and for the transformed Hamiltonian were found; the treatment is also needlessly complicated. Coulter's dressing transformation[19] for an atom within a bounded volume also attempts to satisfy unphysical requirements and it is not convergent in the infinite-volume limit.

Our procedure is based on a generalization of a dressing transformation proposed by Faddeev.[12] We transform

$$H = H_0 + \lambda V \tag{7.1}$$

where the coupling constant λ is the small perturbation parameter, into

$$H' = UHU^{-1} = H_0' + V' \tag{7.2}$$

where U is a unitary transformation,

$$U = \exp(iW), \qquad W = W^\dagger \tag{7.3}$$

and

$$V' = \sum_{n \geq 1} \lambda^n V'_n, \qquad W = \sum_{n \geq 1} \lambda^n W_n \tag{7.4}$$

Equating the coefficients of equal powers of λ, we get

$$V_1' = V + i[W_1, H_0] \tag{7.5}$$

$$V_2' = i[W_1, V] + (i^2/2!)[W_1, [W_1, H_0]] + i[W_2, H_0] \tag{7.6}$$

and so on.

We now choose W_1 so that $i[W_1, H_0]$ cancels out [cf. (7.5)] those terms in V that lead to persistent effects, thus eliminating these effects to first order in λ. With this choice for W_1, we next choose W_2 in (7.6) so that the nondiagonal part of V_2' is free from persistent effects (the diagonal part is incorporated into H_0'). In principle, this procedure can be continued up to arbitrarily high order in λ.

The terms that lead to persistent perturbation effects, according to (3.19), are those that lead to Im $\Sigma_\alpha(E) \neq 0$ for nondissipative states. They are readily recognized in the present problems as those terms that give rise to transitions from the ground state (such as the counter-rotating terms in V).

The choice of $W_1, W_2, \ldots$ is clearly not unique. This nonuniqueness is inherent in any dressing transformation: transformations whose effects differ by an arbitrary admixture of transient terms (i.e., terms not conducive to persistent effects) are equivalent according to the above definition. In practice, our choice of $W_1, W_2 \ldots$ is guided by criteria of simplicity: their form is determined by that of the terms we wish to cancel out, with the insertion of suitable energy denominators to obtain the cancellation when the commutator with H_0 is taken.

8. Dressing Transformation for the Hydrogen Atom

For the sake of simplicity, and in order to facilitate the comparison with the usual treatment, we restrict our discussion of the dressing transformation for the hydrogen atom to the dipole approximation. The Hamiltonian of Section 2 is simplified to

$$H = H_0 + H_{I,1} + H_{I,2} + H_{\text{ren}} \tag{8.1}$$

$$H_0 = \sum_{n \geq 1} E_{0n} |n\rangle\langle n| + \sum_{M=-1}^{1} \int_0^\infty k a_M^\dagger(k) a_M(k) dk \tag{8.2}$$

$$H_{I,1} = \sum_{ijM} \lambda_{ijM} |i\rangle\langle j| \int_0^\infty \rho(k)[a_M(k) + a_M^\dagger(k)] k^{1/2} dk \tag{8.3}$$

where $\rho(k)$ is a cutoff factor [usually taken in the form (4.9)] and

$$\lambda_{ijM} = -(1/m)(2\alpha/3\pi)^{1/2} \langle i| p_M |j\rangle \tag{8.4}$$

p_M being the spherical components[13] of **p**;

$$H_{I,2} = \lambda' \sum_{M=-1}^{1} (-1)^M \int_0^\infty dk\, k^{1/2}\rho(k) \int_0^\infty dk'\, k'^{1/2}\rho(k')$$
$$\times [a_M(k)a_{-M}(k') + (-1)^M a_M^\dagger(k)a_M(k') + \text{H. c.}] \tag{8.5}$$

where

$$\lambda' = \alpha/3\pi m \tag{8.6}$$

We have already written $H_{I,2}$ in normally ordered form, subtracting out a constant term. Finally, in order that m be the experimental mass of the electron, the usual nonrelativistic mass-renormalization counterterm H_{ren} is added. To second order (which is as far as we will carry out the calculation), it is given by

$$H_{\text{ren,2}} = (\Delta m/m)\mathbf{p}^2/2m \tag{8.7}$$

$$\Delta m/m = 4\lambda' \int_0^\infty \rho(k)\, dk \tag{8.8}$$

In order to apply the dressing transformation, some slight modifications in (7.6) are required, owing to the fact that, besides terms linear in the coupling constant ($H_{I,1}$), the Hamiltonian also contains quadratic terms ($H_{I,2}$, H_{ren}).

To first order, the counter-rotating terms that are responsible for persistent effects in (8.3) are those in $|i\rangle\langle 1| a_M^\dagger(k)$ and their Hermitian conjugates. To eliminate them, we choose

$$\lambda W_1 = i \sum_{i,M} \lambda_{i1M} \int_0^\infty dk \, \rho(k) k^{1/2} \frac{|i\rangle\langle 1|a_M^\dagger(k)}{E_{01} - E_{0i} - k} + \text{H.c.} \quad (8.9)$$

Applying a similar procedure to second order, we get for the transformed Hamiltonian

$$H' = H_0' + H_1' + H_2' \quad (8.10)$$

where H_1' is given by (8.3) with the counter-rotating terms connected with the ground state subtracted out, and H_2' is a sum of several terms that will not be written out here[8]; H_0', which incorporates the diagonal contributions found in second order, is given by

$$H_0' = \sum_{n \geq 1} E_{0n}' |n\rangle\langle n| + \sum_{M=-1}^{1} \int_0^\infty k a_M^\dagger(k) a_M(k) dk \quad (8.11)$$

where

$$E_{0n}' = E_{0n} + 4\lambda'\langle n|\mathbf{p}^2/2m|n\rangle \int_0^\infty \rho(k) dk, \qquad n \neq 1 \quad (8.12)$$

$$E_{01}' = E_{01} + \mathscr{L}_1 \quad (8.13)$$

and

$$\mathscr{L}_1 = \sum_{n \neq 1, M} \lambda_{n1M}^2 (E_{0n} - E_{01}) \int_0^\infty \frac{\rho(k) dk}{k + E_{0n} - E_{01}} \quad (8.14)$$

When $\rho(k)$ is replaced by (4.9), $\mathscr{L}_1$ becomes identical with the non-relativistic contribution to the ground-state Lamb shift, as computed by Bethe.[20] Thus, for the ground state, the dressing transformation cancels out the contribution from the mass-renormalization counterterm [cf. (8.12)], and it adds the Lamb-shift correction. Furthermore, the interaction terms H_1' and H_2' in (8.10) produce no persistent effects to second order. We can therefore apply the resolvent operator method, with H' as generator of the time evolution, to discuss decay and resonance fluorescence by procedures similar to those described in Section 6. We confine ourselves to a description of the main results.[8]

The time evolution of an initial excited state $|n\rangle|0\rangle$ is determined, as in Section 6.1, by the analytic properties of $\mathscr{D}_{n;0}(z)$, the associated eigenvalue of the diagonal part of the resolvent. It is found in the second-order treatment that, in analogy with (6.2), the dominant exponentially-decaying term arises from an unphysical-sheet pole

$$z_n = E_n - \tfrac{1}{2} i \Gamma_n = E_{0n} + \mathscr{L}_n - \tfrac{1}{2} i \sum_{j < n, M} \Gamma_{njM} \quad (8.15)$$

where $\mathcal{L}_n$ is the nonrelativistic contribution to the Lamb shift for state n and Γ_{njM} are the partial widths for the transitions to all lower levels.

Thus, the analogue of the energy correction (8.13) for excited states [cancellation of the mass-renormalization counterterm in (8.12) and Lamb-shift correction] appears, as it should, in the poles of the analytic continuation for the resolvent, and not through the dressing transformation, which should lead to the correct energy only for the ground state. This essential difference between ground and excited states was obscured in previous treatments where dressing transformations were attempted.

The discussion of resonance fluorescence also proceeds in analogy with Section 6.2. In order to treat the line-shape problem, we consider, in particular, the differential cross section for scattering of a photon with momentum $\mathbf{k}$ and circular polarization λ from an atom in the ground state, leading to a photon with momentum $\mathbf{k}'$ and circular polarization λ'. The result,[8] obtained with the help of (3.22), is given by a modified Kramers–Heisenberg dispersion formula,

$$\frac{d\sigma}{d\Omega} = r_0^2 \left| \hat{\boldsymbol{\epsilon}}_{\mathbf{k}\lambda} \cdot \hat{\boldsymbol{\epsilon}}^*_{\mathbf{k}'\lambda'} + \frac{1}{m} \sum_{n \neq 1} \frac{\langle 1 | \mathbf{p} \cdot \hat{\boldsymbol{\epsilon}}_{\mathbf{k}\lambda} | n \rangle \langle n | \mathbf{p} \cdot \hat{\boldsymbol{\epsilon}}^*_{\mathbf{k}'\lambda'} | 1 \rangle}{E_{01} - E_{0n} - k} \right.$$

$$\left. + \frac{1}{m} \sum_{n \neq 1} \frac{\langle 1 | \mathbf{p} \cdot \hat{\boldsymbol{\epsilon}}^*_{\mathbf{k}'\lambda'} | n \rangle \langle n | \mathbf{p} \cdot \hat{\boldsymbol{\epsilon}}_{\mathbf{k}\lambda} | 1 \rangle}{E'_{01} + k - [E'_{0n} + \Delta_n(E_{01} + k)] + \tfrac{1}{2} i \Gamma_n(E_{01} + k)} \right|^2 \quad (8.16)$$

where r_0 is the classical electron radius, $\hat{\boldsymbol{\epsilon}}_{\mathbf{k}\lambda}$ and $\hat{\boldsymbol{\epsilon}}_{\mathbf{k}'\lambda'}$ are polarization vectors of the incident and scattered photons, and Δ_n and Γ_n are defined in terms of the resolvent by (3.11). The first line of (8.16) contains the effects of Thomson scattering and of antiresonant terms. If the incident photon energy approaches a resonance associated with a given level n, the corrections Δ_n and Γ_n to the Kramers–Heisenberg formula in the corresponding term of the second summation (8.16) become important. In this case, $E_{0n} + \Delta_n - \tfrac{1}{2} i \Gamma_n$ is close to z_n [cf. (8.15)], so that the Lamb-shift correction and the linewidth are also properly taken into account in the resonance scattering cross section.

9. Interaction Hamiltonian and Line Shape

We finally discuss the connection between the line shape and the choice of the interaction Hamiltonian. The problem is to compare the results obtained with the minimal-coupling interaction Hamiltonian employed in (2.1), in dipole approximation, with those obtained from the interaction Hamiltonian

$$\bar{H}_I = -e\mathbf{r} \cdot \mathbf{E}^+(0) \quad (9.1)$$

which has been widely employed in quantum optics. The relation between these two Hamiltonians seems to have been first discussed by Göppert-Mayer[21] in a semiclassical context, and by Power and Zienau[22] for a quantized field; other recent discussions include that of Woolley.[23] The resonant term obtained from the two different interaction Hamiltonians yields different results for the line shape, and it was remarked by Lamb[2] that only the resonant term derived from (9.1) is in agreement with experiment for the Lamb-shift transition.

In Woolley's treatment, $\mathbf{E}^{\perp} = -\partial\mathbf{A}/\partial t$ is indeed the transverse electric field, while in Power and Zienau's treatment it is replaced by

$$\mathbf{D}^{\perp} = (\mathbf{E} + 4\pi\mathbf{P})^{\perp} \tag{9.2}$$

where

$$\mathbf{P} = e\mathbf{r}\delta(\mathbf{q}) \tag{9.3}$$

is the polarization operator in dipole approximation.

Both interpretations are possible, and the difference between them simply corresponds to regarding the transformation connecting the two Hamiltonians from the active or from the passive point of view.[8]

In the active point of view,[23] which will be adopted here, the new Hamiltonian $\bar{H}$ is connected with H [as given by (2.1) in dipole approximation] by

$$\bar{H} = \exp(-i\Sigma)H\exp(i\Sigma) \tag{9.4}$$

where[24]

$$\Sigma = e\mathbf{r}\cdot\mathbf{A}(\mathbf{0}) \tag{9.5}$$

leading to

$$\bar{H} = \frac{\mathbf{p}^2}{2m} - \frac{e^2}{r} + \frac{1}{8\pi}\int d^3r\,[(\mathbf{E}^{\perp})^2 + (\nabla\times\mathbf{A})^2]$$
$$- e\mathbf{r}\cdot\mathbf{E}^{\perp}(\mathbf{0}) + 2\pi\int[\mathbf{P}^{\perp}(\mathbf{q})]^2d^3q \tag{9.6}$$

where the last term contributes to second-order (or higher-order) calculations, e.g., in the evaluation of the Lamb shift. Here, $\bar{H}$ is regarded as a new Hamiltonian, expressed in terms of the old canonical variables, whereas, in the passive point of view,[22] we would get the old Hamiltonian expressed in terms of new canonical variables.

Since the two Hamiltonians are connected by a unitary transformation, they must lead to equivalent results, provided that the state vectors are correspondingly transformed. If one tries to define the natural line shape in terms of the decay of a specified Weisskopf–Wigner initial state vector, *without* taking into account the transformation of this state vector (it can be shown[8] that this corresponds to adopting different definitions of the photon vacuum), the two Hamiltonians lead to different results, but one cannot say *a priori* which (if any) is to be preferred, because this depends on how realistic it is to regard the associated Weisskopf–Wigner state as being produced by the excitation process. An unambiguous definition of the line shape must include an account of the excitation process.

If we define the line shape in terms of resonance fluorescence, we must discuss the effect of the transformation (9.4) on the S-matrix. The exact S-matrix elements for transitions between corresponding physical states are indeed identical for H and $\bar{H}$. However, when they are defined in terms of the usual adiabatic hypothesis, this is true only after wave function renormalization,[8] and the corresponding renormalization constants Z and $\bar{Z}$ (which represent the probability of finding the unperturbed ground state in the interacting ground state) are different. Since Z and $\bar{Z}$ differ from unity only by terms of order e^2, the Kramers–Heisenberg matrix element is the same for both Hamiltonians, as is well known.[25] However, this need not apply to (8.16), which already includes partial summations over higher-order terms, embodied in the Δ_n and Γ_n corrections. It can be shown,[8] nevertheless, that for photon energies k within the resonance width associated with each given resonance denominator, these corrections also are the same, up to order e^2.

Let r be the resonant level, R the corresponding resonant term [the term $n = r$ in the second line of (8.16)], and B the background, i.e., the sum of all remaining terms in (8.16). The result just stated then implies that, to order e^2,

$$d\sigma/d\Omega = r_0^2 |R + B|^2 = r_0^2 |\bar{R} + \bar{B}|^2 \qquad (9.7)$$

However, we have $R \neq \bar{R}$, $B \neq \bar{B}$. If, as is often done, one approximates the result by retaining only the resonant term, the results are indeed different, and the only way to find out which is the better approximation (apart from comparison with experiment) is to estimate the effect of the background terms.

Let us do this first for the Lyman-α line. In this case, we have

$$\bar{R}/R = k^2/(E_{02} - E_{01})^2 \qquad (9.8)$$

where E_{02} is the energy of the $2p$ level. To compare the line shapes, we characterize them by two parameters: the photon energy k_0 at the peak of the curve and the asymmetry δ, defined by

$$\delta = \left[\left(\frac{d\sigma}{d\Omega} \right)_{k_0 + \Gamma/2} - \left(\frac{d\sigma}{d\Omega} \right)_{k_0 - \Gamma/2} \right] \Big/ \left(\frac{d\sigma}{d\Omega} \right)_{k_0} \qquad (9.9)$$

where Γ is the linewidth associated with the $2p$ level.

In terms of these parameters, we find that

$$(\bar{k}_0 - k_0)/\Gamma = \tfrac{1}{2}\Gamma/k_0 \tag{9.10}$$

and that R is symmetric ($\delta = 0$), whereas for $\bar{R}$,

$$\delta = 2\Gamma/k_0 \tag{9.11}$$

Since $\Gamma/k_0 \approx 4 \times 10^{-8}$, both deviations are extremely small in this case. It may still be asked, however, which one yields a better approximation.

In order to find out, one must compute B, the sum of all "background" terms. This can be done in closed form, with the help of the Coulomb Green's function, which has been employed by Gavrila[26] to compute the Kramers–Heisenberg matrix element.

The result[8] for the photon energy k_0' at the peak when both R and B are included is

$$(k_0' - k_0)/\Gamma \approx -0.22\ \Gamma/k_0 \tag{9.12}$$

and the corresponding asymmetry parameter is

$$\delta' \approx -1.8\ \Gamma/k_0 \tag{9.13}$$

Comparing these results with (9.10) and (9.11), we see that the corrections associated with $\bar{R}$ have the wrong sign, so that, in this case, it is R that represents a better approximation.

As a final example, let us discuss the line shape for the Lamb-shift transition, i.e., for the induced decay from the metastable $2s_{1/2}$ state to the $2p_{1/2}$ state, in the presence of near-resonant microwave photons of frequency k_0 (followed by a Lyman-α transition to the ground state). This corresponds to a more recent version[27] of Lamb's experiment, which does not employ magnetic field tuning of the $2s_{1/2} - 2p_{1/2}$ energy difference.

We make use of the fact that, up to second order, Im $\Sigma_{2s} = 0$ [cf. (3.10)], so that the metastable $2s_{1/2}$ state may be treated as stable, to this order. However, to the same order, Re $\Sigma_{2s} \neq 0$, so that one must correct the energy of the $2s_{1/2}$ state in order for it to behave as an asymptotically stationary state. This is achieved by adding a term to the unperturbed Hamiltonian (8.2) and then subtracting the same term from the interaction Hamiltonian. This term should be nondiagonal, so as to remove the degeneracy between the $2s_{1/2}$ and $2p_{1/2}$ states,[28] and it must not affect the ground-state energy. A suitable choice is[8]

$$H_L = 8\pi\epsilon_0 a_B^3\delta(\mathbf{r}) - 8\epsilon_0|1\rangle\langle 1| \tag{9.14}$$

where a_B is the Bohr radius, and ϵ_0 is the (unrenormalized) nonrelativistic contribution to the Lamb shift of the $2s_{1/2}$ state.

This leads to the replacement

$$E_{02s} \rightarrow E_{02s} + \epsilon \qquad (9.15)$$

in (8.12), where ϵ is the renormalized nonrelativistic contribution to the Lamb shift of the $2s_{1/2}$ state, whereas (8.13) remains unaffected. Corresponding modifications must be made in the dressing transformation to ensure that the $2s_{1/2}$ state (as well as the ground state) remains asymptotically stationary to second order.

A contact term similar to that in (9.14) appears in the usual treatment[29] of the nonrelativistic contribution to the Lamb shift. It has also been employed by Fried[30] as a phenomenological term in a treatment of the same problem based upon a semiclassical Hamiltonian.

Taking the above modifications into account, we find expressions[8] for the differential cross section to second order that can be analyzed in terms of resonant and background contributions, as in (9.7). We find that R is peaked at a photon energy

$$k_0 \approx \epsilon - \Gamma^2/8\epsilon \qquad (9.16)$$

whereas $\bar{R}$ is peaked at

$$\bar{k}_0 \approx \epsilon + \Gamma^2/8\epsilon \qquad (9.17)$$

while the corresponding asymmetries (9.9) are given by

$$\delta \approx -\Gamma/2\epsilon, \qquad \bar{\delta} \approx \Gamma/2\epsilon \qquad (9.18)$$

Since $\Gamma/\epsilon \approx 10^{-1}$, we have $(\bar{k}_0 - k_0)/\epsilon \approx 2.5 \times 10^{-3}$, which is about three orders of magnitude larger than the accuracy of present-day measurements of the Lamb shift. Thus, the differences between R and $\bar{R}$ are easily detectable in this case.

The background contribution B can again be evaluated in closed form,[31] with the help of the Coulomb Green's function. The results for k_0' and δ' when both R and B are included[8] agree with $\bar{k}_0$ and $\bar{\delta}$ of (9.17) and (9.18) up to terms of the order of $\epsilon(\Gamma/\epsilon)^4$ and $(\Gamma/\epsilon)^2$, respectively. This remains valid when fine- and hyperfine-structure contributions are taken into account.

Thus, in agreement with Lamb's remark,[2] the resonant term derived from (9.1) indeed yields better results for the line shape in the Lamb-shift transition. However, this does not hold true in other cases, as shown by our discussion of the Lyman-α line. Similar conclusions were reached by Fried.[30]

The line shape in the Lamb-shift transition, defined in terms of the usual minimal-coupling Hamiltonian of quantum electrodynamics, illustrates the need to go beyond the Weisskopf–Wigner approximation. In order to obtain agreement with experiment, one must take into account the effect of (virtual) transitions to all nonresonant levels. These background corrections to the line shape can be systematically evaluated by employing the Coulomb Green's function.

As the accuracy of Lamb-shift measurements increases, the evaluation of background corrections to the line shape becomes of comparable or possibly greater significance than that of higher-order radiative corrections to the resonant term, since the result of the measurement is directly affected by the line shape, which arises from the interference between resonant and background contributions.

References and Notes

1. V. Weisskopf and E. P. Wigner, *Z. Phys.* **63**, 54 (1930).
2. W. Lamb, *Phys. Rev.* **85**, 259 (1952).
3. W. Heitler and S. T. Ma, *Proc. R. Ir. Acad. Sect. A* **52**, 109 (1949); E. Arnous and S. Zienau, *Helv. Phys. Acta* **24**, 279 (1951); E. Arnous and K. Bleuler, *Helv. Phys. Acta* **25**, 581, 631 (1952).
4. E. Arnous and W. Heitler, *Proc. R. Soc. London Ser. A* **220**, 290 (1953).
5. W. Heitler, *The Quantum Theory of Radiation*, 3rd ed., Oxford University Press, London (1954).
6. F. E. Low, *Phys. Rev.* **88**, 53 (1952).
7. H. M. Nussenzveig, *Causality and Dispersion Relations*, Academic, New York (1972), Chapter 4.
8. L. Davidovich, Ph.D. thesis, University of Rochester (1975); L. Davidovich and H. M. Nussenzveig, to be published.
9. H. E. Moses, *Phys. Rev. A* **8**, 1710 (1973).
10. L. Van Hove, *Physica (Utrecht)* **21**, 901 (1955).
11. For a historical survey, see G. S. Agarwal, *Quantum Statistical Theories of Spontaneous Emission*, Springer-Verlag, Berlin (1974).
12. L. D. Faddeev, *Sov. Phys. Dokl.* **8**, 881 (1964).
13. M. E. Rose, *Elementary Theory of Angular Momentum*, Wiley, New York (1957).
14. F. J. Dyson, *Phys. Rev.* **75**, 486, 1736 (1949).
15. E. Grimm and V. Ernst, *J. Phys. A: Gen. Phys.* **7**, 1664 (1974); *Z. Phys. A* **274**, 293 (1975).
16. E. B. Davies, *J. Math Phys.* **15**, 2036 (1974).
17. See E. Grimm and V. Ernst, *Z. Phys. A.* **274**, 293 (1975) for an example in which this condition is not fulfilled.
18. E. C. G. Stueckelberg and D. Rivier, *Helv. Phys. Acta* **23**, 215 (1950); M. Fierz, *Helv. Phys. Acta* **23**, 731 (1950).
19. C. A. Coulter, *Phys. Rev. A.* **10**, 1946 (1974).
20. H. A. Bethe, *Phys. Rev.* **72**, 339 (1947).
21. M. Göppert-Mayer, *Ann. Phys. (Leipzig)* **9**, 273 (1931).
22. E. A. Power and S. Zienau, *Nuovo Cimento* **6**, 7 (1957), *Philos. Trans. R. Soc. London Ser. A* **251**, 427 (1959); see also M. Babiker, E. A. Power, and T. Thirunamachandran, *Proc. R. Soc. London Ser. A* **338**, 235 (1974).
23. R. G. Woolley, *Mol. Phys.* **22**, 1013 (1971).
24. A cutoff in momentum space is required in order for (9.4) and (9.5) to define a proper unitary transformation. One can adopt the same cutoff as in (8.3).

25. P. A. M. Dirac, *The Principles of Quantum Mechanics,* 4th ed., Oxford University Press, London (1958).
26. M. Gavrila, *Phys. Rev.* **163,** 147 (1967).
27. S. R. Lundeen and F. M. Pipkin, *Phys. Rev. Lett.* **34,** 1368 (1975).
28. If this degeneracy is not removed, the first-order contribution of the minimal-coupling interaction to the $2s_{1/2} \to 2p_{1/2}$ + one photon transition vanishes, and one must compute higher-order contributions, leading essentially to the same results [see E. J. Kelsey, *Phys. Rev. A* **15,** 647 (1977)].
29. J. J. Sakurai, *Advanced Quantum Mechanics,* Addison-Wesley, Reading, Massachusetts (1967).
30. Z. Fried, *Phys. Rev. A* **8,** 2835 (1973).
31. S. Klarsfeld, *Lett. Nuovo Cimento* **1,** 682 (1969).

8

Resonance Fluorescence and Spontaneous Emission as Tests of QED

K. Wódkiewicz

Do optical photons exist? Should we quantize the electromagnetic field in order to describe effects in the optical region? For more than twenty years these questions have been discussed in the domain of optical phenomena. Although QED provides a correct picture of all phenomena in the optical region, it has been argued that the photoelectric effect or the theory of spontaneous emission can be treated semiclassically and the quantum nature of the electromagnetic field is not necessary to describe these effects.

Because a surprisingly large number of physical effects in the optical region can be explained without field quantization, a constant search for a pure quantum feature of the electromagnetic field has been made in order to exhibit the validity and the necessity of QED in low-frequency phenomena.[1,2]

For the purpose of this paper, I shall restrict the discussion of the quantum properties of radiation theory to only two physical phenomena: spontaneous emission from a single atom and spontaneous emission in the presence of a strong driving field. The last effect is known as resonance fluorescence and has been very carefully examined both theoretically and experimentally in the last few years.[3]

I shall emphasize those properties of spontaneous emission and resonance fluorescence that exhibit quantum features of the electromagnetic field and that cannot be explained on semiclassical grounds even with some dynamical modifications such as radiation reaction.[4] In order to make contact with experiment and to avoid introducing notions specific

K. Wódkiewicz • Institute of Theoretical Physics, Warsaw University, Warsaw 00-681, Poland

to any particular formulation of radiation theory, I shall further restrict consideration of the properties of spontaneous emission and resonance fluorescence to those that can be obtained by photodetection correlation measurements.

A systematic investigation of photocounting experiments permits a detailed analysis of the correlation properties of the radiating system. In order to have a full understanding of the statistical properties of the electromagnetic field we have to know in general an infinite number of the so-called coherence functions of the electric field[5]

$$G^{n,m}(t_1, \cdots, t_n; t_{n+1}, \cdots, t_{n+m})$$
$$= \langle E^{(-)}(t_1) \cdots E^{(-)}(t_n) E^{(+)}(t_{n+1}) \cdots E^{(+)}(t_{n+m}) \rangle \quad (1)$$

All interesting physical quantities can be related to proper coherence functions. The intensity $I(t)$ of the scattered radiation is simply equal to $G^{1,1}(t;t)$. The two-point correlation function $G^{1,1}(t; t + \tau)$ contains information about the power spectrum, the stationarity or the nonstationarity, and the proper coherence times of the emitted light. For stationary correlations, the power spectrum $P(\omega)$ of the fluorescence is simply a Fourier transform of $G^{1,1}(t; t + \tau)$ with respect to τ. Another very important measurable quantity is the joint probability of photodetection $P_2(t; t + \tau)$ by a phototube at time $t + \tau$, if photodetection took place at time t. It is well known that this joint probability is proportional to the fourth-order correlation function $G^{2,2}(t + \tau, t; t + \tau, \tau)$.[6] These definitions do not depend on whether the electromagnetic field is quantized or not. The only difference lies in the understanding of the right-hand side of the relation given by equation (1), where the expectation value $\langle \cdots \rangle$ for a classical case should be understood as a stochastic average over an ensemble of classical fields if a stochastic approach is used, or should be simply dropped for a classical theory where the electric field can be derived from some dynamical equations of motion. A stochastic expectation value is also understood for random electrodynamics with a universal, randomly fluctuating zero-point field.[7]

For QED the particular form of the expectation value depends on the formulation of the theory. One can use operators, Green's functions, or Feynman path integration. Whatever technique is applied, the net result is that QED predicts for spontaneous emission and resonance fluorescence a set of correlation functions different from that of the semiclassical theory.

Now the situation regarding a search for quantum features in spontaneous emission and resonance fluorescence is clear. We shall look for coherence functions, i.e., experimental situations, for which the semiclassical theory and all other alternative theories lead to results different from those of QED. What is very important to remember is the fact that one needs an infinite set of correlation functions in order to fully describe the statistical properties of radiation. One cannot base any solid argument for or against QED on only one coherence function. In this lecture I will show that higher correlation functions for spontaneous emission and resonance fluorescence are equally important in order to draw proper conclusions about the quantum nature of the electromagnetic field.

In many textbooks on elementary quantum mechanics and radiation theory, it is written that obvious evidence for the quantum nature of the electromagnetic field is found in the decay rate of a spontaneously radiating atom, i.e., $I(t) = I_0 e^{-At}$, where A is the famous Einstein coefficient of spontaneous emission. This statement really means that the behavior of the function $G^{1,1}(t; t)$ rules out a nonquantized theory of radiation. Let me recall now a historical example showing that one should be very careful with such statements.

In a little-known paper written in 1927, Fermi, using only classical electrodynamics and the radiation principle combined with the Schrödinger interpretation of the wave function, proposed the following nonlinear generalization of the Schrödinger equation[8]:

$$i\hbar(\partial\psi/\partial t) = [-(\hbar^2/2m)\Delta + V]\psi + V_{RR}\psi \qquad (2)$$

where

$$V_{RR}(\psi) = -(2e/3c^3)\,\ddot{\mathbf{d}}\cdot\mathbf{r} \qquad (3)$$

is the radiation reaction potential with

$$\mathbf{d} = e\int d^3r\,\psi^*(\mathbf{r})\mathbf{r}\psi(\mathbf{r})$$

being the dipole moment of the radiating charge according to the Schrödinger interpretation of the wave function. The main goal of Fermi's generalization of the Schrödinger equation was to understand the mechanism of radiation due to the charge distribution of the wave function in an atom. If only two atomic levels with energy E_2 and E_1 are involved in the problem, we can write the wave function of the system in the following form:

$$\psi = a_1 \exp(-iE_1t/\hbar)\varphi_{E_1} + a_2 \exp(-iE_2t/\hbar)\varphi_{E_2} \qquad (4)$$

with $[-(\hbar^2/2m)\Delta + V]\varphi_{E_{1,2}} = E_{1,2}\varphi_{E_{1,2}}$. According to Fermi, equation (2) leads to the following set of nonlinear equations (secular contribution only) for the coefficients a_1 and a_2:

$$\dot{a}_1 = (A/2)a_1 a_2^* a_2 \qquad (5a)$$

$$\dot{a}_2 = -(A/2)a_2 a_1^* a_1 \qquad (5b)$$

with

$$A = (4\, e^2/3\, c^3\hbar)\, [E_2 - E_1)^3/\hbar^3] \left(\int \varphi_{E_1}{}^*\mathbf{r}\; \varphi_{E_2} d^3 r \right)^2$$

The nonlinear equations (5) can be solved exactly with arbitrary initial conditions:

$$|a_1(t)|^2 = |a_1(0)|^2 \frac{e^{At}}{1 - |a_1(0)|^2 + e^{At}|a_1(0)|^2} \tag{6a}$$

$$|a_2(t)|^2 = \frac{1 - |a_1(0)|^2}{1 - |a_1(0)|^2 + e^{At}|a_1(0)|^2} \tag{6b}$$

If the initial state is an energy eigenstate, i.e., $a_2(0) = 1$ and $a_1(0) = 0$, we see immediately that no decay can occur. If $|a_2(0)|^2 = 1 - \epsilon$, where ϵ is an arbitrary small number, for $At \gg 1$ we obtain $|a_2(t)|^2 \propto e^{-At}$, i.e., the excited state decays with a rate equal to the Einstein A coefficient. It turns out that the numerical value of A agrees with the one obtained by Dirac in his famous paper on the quantization of the electromagnetic field.[9]

From the structure of the solutions (6) it is clear why Fermi did not develop this theory further. The nonlinear Schrödinger equation with the radiation reaction potential (3) leads to very serious problems with the time behavior of spontaneous emission. An exact eigenstate ($\epsilon = 0$) cannot radiate spontaneously and even if $\epsilon \neq 0$ it can take a very long time for the atom to get even halfway deexcited. Regardless of these fundamental difficulties, a theory based on the Fermi idea has been advocated for years by Jaynes and co-workers as the neoclassical theory of radiation.[4]

This historical example shows that in discussions of the quantum properties of radiation, the spontaneous emission rate or even the radiative level shifts cannot be regarded as crucial tests of QED. This fact, based on the properties of the function $G^{1,1}(t;\, t)$, has sometimes been stated much more strongly by saying that spontaneous emission is an inadequate test of QED.[10] It has been argued next that spontaneous emission from a strongly driven atom by a monochromatic source of light offers a much better test of the quantum properties of the radiation field and recent work has concentrated on the use of resonance fluorescence as a test of QED in the optical region. As it has been said before, this statement was based on the properties of the function $G^{1,1}(t;\, t)$. We shall show that spontaneous emission and resonance fluorescence contain equally the properties of the quantum electromagnetic field if one looks at higher coherence functions and that these properties cannot be obtained from any of the neoclassical or semiclassical arguments.

We start our discussion with the following semiclassical Bloch equations describing a model two-level atom coupled to a *c*-number electromagnetic field $\mathscr{A}(t)$ in the *rotating-wave approximation* (RWA)[11]:

$$\dot{\sigma} = - i\omega_0 \sigma - (1/2\tau_0) \sigma + g \sigma_3 \mathscr{A}(t) \tag{7a}$$

$$\dot{\sigma}_3 = - (1/\tau_0)(1 + \sigma_3) - 2g [\sigma^+ \mathscr{A}(t) + \mathscr{A}^*(t)\sigma] \tag{7b}$$

where $\sigma = |1\rangle\langle 2|$ is the atomic transition operator and $\sigma_3 = |2\rangle\langle 2| - |1\rangle\langle 1|$ is the atomic population inversion. The strength of the coupling between the atomic system and the *c*-number field is described by the coupling constant *g*. In equations (7) additional damping terms have been added phenomenologically in order to take into account spontaneous emission of the two-level atom in the absence of the external field ($\mathscr{A} = 0$). It is assumed that $1/\tau_0$ is equal to the Einstein *A* coefficient. According to classical electrodynamics the electric field in the far zone from a radiating atom is proportional to the dipole moment of the radiating charge. In the case of a two-level atom we have simply the following relations:

$$\mathscr{E}(t) \propto \langle\sigma(t)\rangle/r \qquad \mathscr{E}^*(t) \propto \langle\sigma^+(t)\rangle/r \tag{8}$$

i.e., the emitted light is proportional to the quantum expectation values of the two-level dipole moment operators. The expectation values in equation (8) are computed with respect to the quantum state of the atomic system and as such have nothing to do with the statistical properties of the electromagnetic field.

A theory with the quantized electromagnetic field leads to equations of motion which look formally like the semiclassical equations (7). As was shown by Eberly in Chapter 2 of this volume, QED of a two-level atom coupled to the radiation field leads to the following Heisenberg equations of motion (in the Markovian limit):

$$\dot{\sigma} = - i\omega_0 \sigma - (1/2\tau_0) \sigma + g \sigma_3 A^{(+)}(t) \tag{9a}$$

$$\dot{\sigma}_3 = - [(1 + \sigma_3)/\tau_0] - 2g [\sigma^+ A^{(+)}(t) + A^{(-)}(t)\sigma] \tag{9b}$$

There are however two fundamental differences between the theories given by equations (7) and equations (9). First the *c*-number fields $\mathscr{A}(t)$ and $\mathscr{A}^*(t)$ have been replaced in equations (9) by the annihilation $A^{(+)}(t)$ and the creation $A^{(-)}(t)$ operators of the free electromagnetic field. The next difference, which is a consequence of the quantized dynamics of the electromagnetic field, lies in the definition of the emitted radiation [see equation (8)]. Quantum electrodynamics leads to the following definitions of the emitted electric field operators:

$$E^{(+)}(t) \propto \sigma(t)/r \qquad E^{(-)}(t) \propto \sigma^+(t)/r \tag{10}$$

Let us consider first the power spectrum of the emitted light in spontaneous emission and in strong-field resonance fluorescence. As was pointed out, the power spectrum is given by the following formula:

$$P(\omega) = 2 \operatorname{Re} \int_0^\infty e^{i\omega\tau} G^{1,1}(t + \tau; t) \tag{11}$$

where

$$G^{1,1}(t + \tau; t) = \begin{cases} \mathscr{E}^*(t + \tau)\,\mathscr{E}(t) & \text{for semiclassical case} \\ \langle E^{(-)}(t + \tau)E^{(+)}(t)\rangle & \text{for QED} \end{cases} \tag{12}$$

The QED expression requires an average over the atomic and field states of the system. According to QED the proper electric field correlation function required in expression (11) can be computed exactly from equations (9) with the initial state of the electromagnetic field being the photon-vacuum state:

$$\langle E^{(-)}(t + \tau)E^{(+)}(t)\rangle \propto \exp\left(-t/2\tau_0\right)$$

$$\times \operatorname{Tr}[\rho_A \sigma^+(0)\sigma(0)] \exp[-i\omega_0\tau - (\tau/2\tau_0)] \tag{13}$$

We observe first that this correlation function is not stationary and depends on time through the damping factor $\exp(-t/2\tau_0)$. This nonstationarity is rather trivial and can be handled simply by saying that if we correlate the outgoing electric field at different times $t + \tau$ and t, the result should vanish for $t \gg \tau_0$ because the atom has already emitted spontaneously. Only correlations for $t \gtrsim \tau_0$ can give significant contributions to the power spectrum. In this limit equations (11) and (13) lead to the well-known Lorentzian shape with width equal to $1/\tau_0$. The semiclassical equations of motion for spontaneous emission lead to the following expression for the outgoing electric field:

$$\mathscr{E}(t) \propto (\langle\sigma(0)\rangle/r)\exp[-i\omega_0 t - (t/2\tau_0)] \tag{14}$$

where $\langle\sigma(0)\rangle = \operatorname{Tr}[\rho_A\sigma(0)]$ and ρ_A is the density matrix of the two-level atom. It is clear that for an atom in its upper excited state $\langle\sigma(0)\rangle = 0$, and the spectrum would vanish if only semiclassical arguments were applied. Thus, even with phenomenological decay constants the semiclassical equations of motion (7) fail to describe properly the spectrum of spontaneous emission of an excited two-level system. Both QED and the semiclassical theory predict the right behavior of the atomic dipole operator in spontaneous emission:

$$\langle\sigma(t)\rangle = \langle\sigma(0)\rangle \exp[-i\omega_0 t - (t/2\tau_0)] \tag{15}$$

but conclusions about the spectrum are different. Thus, any argument about quantum properties of the radiation field based on $\langle\sigma(t)\rangle$ would lead to inconclusive statements. The higher-order correlation function $G^{1,1}(t + \tau; t)$ shows clearly that the semiclassical theory fails to describe the spontaneous line shape.

In order to describe resonance fluorescence we assume that the external field in the semiclassical equations (7a) and (7b) has the following form $\mathscr{A}(t) = \mathscr{A}_0 \exp(-i\omega_0 t)$ (exact resonance of the atom and the driving field). The same assumption for the QED Bloch equations (9a) and (9b) is equivalent to a choice of the initial state of the electromagnetic field as a single-mode coherent state: $A^{(+)}|\mathscr{A}_0 \exp(-i\omega_0 t)\rangle = \mathscr{A}_0 \exp(-i\omega_0 t) \times |\mathscr{A}_0 \exp(-i\omega_0 t)\rangle$. It is easy to check now that both equations (7) and (9) lead to the following solution for the atomic dipole operator:

$$\langle \sigma(t) \rangle = \exp(-i\omega_0 t) \{a_0 + a_+ \exp[-(3t/4\tau_0) + i\Omega t] + a_- \exp[-3t/4\tau_0 - i\Omega t]\} \quad (16)$$

where the form of the constants $a_0, a_\pm$ is irrelevant for our further considerations and $\Omega^2 = 4g|\mathscr{A}_0|^2$ is the Rabi frequency. Now it is clear from the form of the solution given by equation (16) that the semiclassical power spectrum of resonance fluorescence consists of two sidebands located at a distance Ω from a sharp contribution centered at ω_0. This means that the observed three-peaked spectrum of the emitted light cannot be explained by a simple Fourier transform of the radiation dipole expectation value [equation (16)].[12] QED requires the computation of the correlation function $\langle E^{(-)}(t + \tau) E^{(+)}(t) \rangle$ in order to obtain the power spectrum of the fluorescent light. Again as in the case of spontaneous emission this correlation function is a nonstationary function of t and τ. Due to a persistent driving field the nonstationarity in the strong-field resonance fluorescence has a highly nontrivial behavior which is a result of the quantum properties of the radiation field.[13]

The measured experimental power spectrum both in spontaneous emission and resonance fluorescence indicates that the electric field correlation function does not factorize itself, i.e., contradicts the semiclassical approach. The QED calculations of the power spectrum lead to an excellent agreement with very accurate experiments.

Let us consider now a photon correlation experiment in which $P_2(t; t + \tau)$ is measured. For stationary processes the proper fourth-order correlation function can be written as follows:

$$G^{2,2}(t; t + \tau) = G^{1,1}(t; t)\, G^{1,1}(t + \tau; t + \tau)\, [1 + \lambda(\tau)] \quad (17)$$

For classical or stochastic electromagnetic fields one can show that we should always have $\lambda \geq 0$. States with no classical analog lead to $\lambda < 0$. Such states of the electromagnetic field for which $\lambda(\tau)$ is negative lead to the so-called photon-antibunching effect.

Let us consider now the problem of photon antibunching in both resonance fluorescence and spontaneous emission. In reported experiments the phenomenon of antibunching has been observed for a single atom coupled to a continuous-wave (cw) dye laser.[14] In these experiments the measured probability of photodetection $P_2(t; t + \tau)$ at two times t and $t + \tau$ was found to be an increasing function of τ from $\tau = 0$. This effect has no classical interpretation and it is believed that its observation would be an explicit feature of the quantized electromagnetic field. Moreover, it is argued in Reference 14 that the antibunching in resonance fluorescence is rather direct evidence for an atom undergoing a quantum jump and for the existence of optical photons. The experiment shows that the joint probability $P_2(t; t + \tau)$ vanishes for $\tau = 0$. This indicates that the atom, having emitted a photon at time t, cannot radiate again immediately. It can radiate again only after being driven back to the upper state by the cw laser. The population of the upper state is modulated at the Rabi frequency, which in the experiment was larger than the natural decay rate of the sodium excited state. These modulations have also been observed.

According to the definition of the joint probability of photodetection we have the following relations:

$$P_2(t; t + \tau) \propto \begin{cases} \langle \mathscr{E}^*(t)\mathscr{E}^*(t + \tau)\mathscr{E}(t + \tau)\mathscr{E}(t) \rangle & \text{(18a)} \\ \langle E^{(-)}(t)E^{(-)}(t + \tau)E^{(+)}(t + \tau)E^{(+)}(t) \rangle & \text{(18b)} \end{cases}$$

for the semiclassical case and for the quantized case, respectively. The semiclassical theory leads to factorization of the joint probability:

$$P_2(t; t + \tau) = P_1(t)P_1(t + \tau) \tag{19}$$

This means that the function λ in the definition (17) simply is equal to zero which is in agreement with the classical region ($\lambda \geq 0$). A stochastic field would lead to the function $\lambda \geq 0$, so again we are in the classical region. The neoclassical arguments, whatever the details of calculations, would lead to a factorization of the joint probability, i.e., $\lambda = 0$.

Now the QED calculations of the correlation function (17), using Heisenberg equations of motion (9) for a two-level atom undergoing spontaneous emission in the absence of an external field, leads to the following solution:

$$P_2(t; t + \tau) = \exp(-\tau/\tau_0) \langle \sigma^+(t)\sigma^+(t)\sigma(t)\sigma(t) \rangle \tag{20}$$

Due to the algebraic identity $\sigma^2 = 0$, we find that the joint probability $P_2(t; t + \tau)$ is equal to zero, clear theoretical evidence for the atom undergoing a quantum jump while emitting a photon spontaneously.[15] It is obvious that in this case $\lambda(\tau) = -1$, in sharp contrast with the semiclassical prediction. This simply means that the state of the interacting system has no classical analog. Thus, the fourth-order correlation function of the field exhibits the same pure quantum features in spontaneous emission as in resonance fluorescence and provides direct evidence for the quantum nature of the electromagnetic field.

The conclusions of our discussion are the following. Both resonance fluorescence and spontaneous emission contain corresponding pure quantum features of the electromagnetic field. In each case these are manifest only if second- and fourth-order correlation functions of the field are investigated. The second-order correlation function describes the spectrum of the emitted light and the fourth-order correlation function provides evidence for the atom performing a quantum jump. In each case these effects are direct results of QED and cannot be explained in the semiclassical approach. Arguments based on rates, i.e., simple expectation values of the field, can lead to inconclusive statements about features of the electromagnetic field.

ACKNOWLEDGMENTS

The author is grateful to Professor J. H. Eberly for many discussions and his hospitality at the University of Rochester where part of this work was done. I am also indebted to Professor I. Bialynicki-Birula, Dr. J. A. Ackerhalt, Dr. P. W. Milonni, and Dr. K. Rzażewski for useful discussion over an extended period of time.

References and Notes

1. See, for example, L. Mandel, *Progress in Optics,* Vol. 13, Ed. E. Wolf, North-Holland, Amsterdam (1975).
2. P. W. Milonni, *Phys. Rep. C* **25,** 1 (1976).
3. See the review papers in *Multiphoton Processes,* Eds. J. H. Eberly and P. Lambropoulos, Wiley, New York (1978).
4. E. T. Jaynes in *Coherence and Quantum Optics,* Eds. L. Mandel and E. Wolf, Plenum, New York (1973).
5. H. M. Nussenzveig, *Introduction to Quantum Optics,* Gordon and Breach, New York (1973).
6. L. Mandel and E. Wolf, *Rev. Mod. Phys.* **37,** 231 (1965).
7. T. H. Boyer, *Phys. Rev. D* **11,** 790, 809 (1975), and Chapter 5 of this volume.
8. E. Fermi, *Rend. Lincei* **5,** 795 (1927).
9. P. A. M. Dirac, *Proc. R. Soc. London Ser. A* **114,** 243 (1927).
10. H. J. Kimble and L. Mandel, *Phys. Rev. Lett.* **34,** 1485 (1975).
11. See, for example, L. Allen and J. H. Eberly, *Optical Resonance and Two-Level Atoms,* Wiley, New York (1975).
12. C. Cohen-Tannoudji, *Proceedings of the Second Laser Spectroscopy Conference,* Megeve, France, 1975, Spring-Verlag, Berlin (1975).
13. J. H. Eberly, C. Kunasz, and K. Wódkiewicz, *J. Phys. B.* (in press).
14. H. J. Kimble, M. Dagenais, and L. Mandel, *Phys. Rev. Lett.* **39,** 691 (1977).
15. J. H. Eberly and K. Wódkiewicz, unpublished.

9

Phase Representation of Intense Photon Beams and Its Applications

Iwo Bialynicki-Birula

1. Introduction

Ever since its introduction by Dirac in his classic paper on the quantum theory of radiation,[1] the phase of the quantized electromagnetic field has been the subject of intensive studies and even some controversies. Extensive reviews of these problems with emphasis on the proper mathematical description have been given by Carruthers and Nieto[2] and by Paul.[3]

In the present paper I shall be concerned with some practical uses of the phase of the quantized electromagnetic field in the study of intense photon beams. Section 2 contains a brief description of the method and in the remaining sections two applications are described. In Section 3 the spectrum of an N-level atomic system interacting in the dipole approximation with a single mode of the quantized electromagnetic field is analyzed. In Section 4 the energy levels of a relativistic electron moving in the field of a quantized plane wave are determined.

2. Phase Representation of Intense Photon Beams

Let us consider a single mode of the quantized electromagnetic field, or for that matter the one-dimensional quantum harmonic oscillator. Following our earlier publications,[4,5] I shall introduce a representation of the eigenstates $|n\rangle$ of the number operator of the system in terms of

Iwo Bialynicki-Birula • Institute of Theoretical Physics, Warsaw University, Warsaw, Poland, and Department of Physics, University of Pittsburgh, Pittsburgh, Pennsylvania

harmonic wave functions of an auxiliary variable φ. For our purposes it is convenient to "shift" the eigenvalues by a constant integer and to adopt the following correspondence rule between the state vectors and wave functions:

$$|n + m\rangle \rightarrow e^{im\varphi} \tag{1}$$

The number n is assumed to be a fixed reference point, while m is varying. Having established the representation of the basic vectors, we may also find the representation of all operators, notably the annihilation operators, the creation operators, and the number operator:

$$a \rightarrow e^{-i\varphi}\left(n + \frac{1}{i}\frac{\partial}{\partial\varphi} \right)^{1/2} \tag{2a}$$

$$a\dagger \rightarrow \left(n + \frac{1}{i}\frac{\partial}{\partial\varphi} \right)^{1/2} e^{i\varphi} \tag{2b}$$

$$a\dagger a \rightarrow n + \frac{1}{i}\frac{\partial}{\partial\varphi} \tag{3}$$

The relations (1), (2), and (3) form the basis of our phase representation of state vectors and operators. Levy-Leblond[6] has independently advocated the usefulness of such a representation.

We can take full advantage of the phase representation in the case of highly populated states when the reference number n is very large. We may then expand the expressions (2) into a series of inverse powers of n and keep only the lowest terms

$$a \sim n^{1/2} e^{-i\varphi}\left(1 + \frac{1}{2in}\frac{\partial}{\partial\varphi} + \cdots \right) \tag{4a}$$

$$a\dagger \rightarrow n^{1/2}\left(1 + \frac{1}{2in}\frac{\partial}{\partial\varphi} + \cdots \right) e^{i\varphi} \tag{4b}$$

In the phase representation, in the limit when $n \rightarrow \infty$, all formulas resemble very much their semiclassical counterparts, while the interpretation of all equations is still fully quantum mechanical.

3. N-Level System Interacting with Quantized Radiation

Let us consider an atomic system or a molecular system with N equally spaced energy levels interacting with a single mode of the quantized electromagnetic field. The Hamiltonian of this system, in the dipole approximation, may be written in the form

$$H = \omega a\dagger a + \omega_0 S_3 + \kappa S_1 (a + a\dagger) \tag{5}$$

where S_i are the standard spin matrices for a spin s ($2s = N - 1$) system. In the phase representation, in the limit when $n \to \infty$, the Hamiltonian (3) can be written as

$$H = n\omega + \omega \frac{1}{i} \frac{\partial}{\partial \varphi} + \omega_0 S_3 + 2\kappa n^{1/2} S_1 \cos \varphi \qquad (6)$$

The energy eigenvalues E are to be determined from the eigenvalue equation

$$\left(n + \frac{1}{i} \frac{\partial}{\partial \varphi} + \mu S_3 + \rho S_1 \cos \varphi \right) \psi_\epsilon(\varphi) = \epsilon \psi_\epsilon(\varphi) \qquad (7)$$

where the following dimensionless variables have been introduced

$$\mu = \omega_0/\omega \qquad (8a)$$

$$\rho = 2\kappa n^{1/2}/\omega \qquad (8b)$$

$$\epsilon = E/\omega \qquad (8c)$$

On account of (1), physically acceptable solutions must obey the periodicity condition:

$$\psi(2\pi) = \psi(0) \qquad (9)$$

which leads to a spectrum of eigenvalues. Equation (7) represents a set of N coupled, first-order differential equations. Its solutions may be formally represented in terms of the Dyson ordering operation:

$$\psi_\epsilon(\varphi) = e^{-i(n-\epsilon)\varphi} U(\varphi) \psi_\epsilon(0) \qquad (10)$$

where

$$U(\varphi) = T \exp\left(-i \int_0^\varphi d\varphi K \right) \qquad (11)$$

and

$$K = \mu S_3 + \rho S_1 \cos \varphi \qquad (12)$$

The periodicity condition (9) leads to the following relation:

$$[e^{2\pi i \epsilon} U(2\pi) - 1] \psi_\epsilon(0) = 0 \qquad (13)$$

Therefore, the determinant of the $N \times N$ matrix appearing in the square bracket must vanish. This will be the case if $e^{-2\pi i \epsilon}$ is equal to one of the eigenvalues of the matrix $U(2\pi)$. The values of ϵ are determined up to an additive constant equal to a natural number. At this point our approach exhibits many formal similarities with the method of quasienergies developed by Zeldovich,[7] but the physical interpretation of the two methods is quite different.

The calculation of $U(\varphi)$ is simplified by noting its group-theoretic content. To see this clearly let us divide the interval $(0,2\pi)$ into m equal subintervals and use the definition of the ordered product to obtain the following approximate expression for $U(2\pi)$;

$$U(2\pi) \simeq \exp(-i\Delta\varphi K_m) \cdots \exp(-i\Delta\varphi K_l) \qquad (14)$$

where

$$K_l = \mu S_3 + \rho \, S_1 \cos \varphi_l \qquad (15)$$

and

$$\varphi_l = 2\pi l/m \qquad (16)$$

Each factor in expression (14) is a unitary operator representing a certain rotation. A succesion of rotations is again a rotation. In the limit, when $m \to \infty$, we have a "continuous sequence" of rotations. Since the matrices S_i are the generators of rotations, the product (14) may always be written in the form

$$\exp(i\boldsymbol{\gamma}\cdot\mathbf{S}) \qquad (17)$$

The vector $\boldsymbol{\gamma}$ represents geometrically the rotation and hence it does not depend on the representation. We may evaluate it using any representation of the matrices S_i. Of course, the simplest case is that of spin $\frac{1}{2}$, and this was the representation used in our numerical calculations. Multiplication of 2×2 matrices is so simple that it can be easily handled with the use of a programmable pocket calculator. The simplest form of the matrix (17) is obtained when the z-axis is chosen in the direction of $\boldsymbol{\gamma}$; this results in the diagonalization of the matrix $\boldsymbol{\gamma}\cdot\mathbf{S}$,

$$\exp(i\boldsymbol{\gamma}\cdot\mathbf{S}) = \mathrm{diag}(e^{i\gamma S}, e^{i\gamma(S-1)}, \ldots, e^{-i\gamma S}) \qquad (18)$$

where γ is the length of $\boldsymbol{\gamma}$. Hence the eigenvalues of the energy operator for an N-level system are

$$E_{n,k} = n\omega - k\omega\gamma/2\pi \qquad 2k = (-N, -N+1, \ldots, N) \qquad (19)$$

Not only the eigenvalues of H, but also the transition probabilities can be easily calculated with the use of the phase representation. For example, the spontaneous or induced transition probabilities involve matrix elements of S_1. Thus we have to evaluate scalar products

$$(\psi_f, S_1\psi_i) \qquad (20)$$

where the initial and final satates are the dressed states of the atomic system, i.e., the vectors $\psi_{f,i}$ are solutions of equation (13). The modulus squared of (20) measures the oscillator strength, i.e., the intensity of the corresponding transition.

4. Dirac Particle Interacting with an Intense Photon Beam

Let us consider an electron, described by the relativistic Dirac theory, interacting with a single mode of the quantized electromagnetic field, characterized by the wave vector **k** and the polarization vector **e**;

$$H = \hbar\omega a^\dagger a + c\boldsymbol{\alpha}\cdot\left[\frac{\hbar}{i}\nabla - \frac{e}{c}\mathbf{A}(\mathbf{r})\right] + \beta mc^2 \qquad (21)$$

where

$$\mathbf{A}(\mathbf{r}) = (\hbar c^2/2\omega V)^{1/2}\,(\mathbf{e}\,a\,e^{i\mathbf{k}\cdot\mathbf{r}} + \mathbf{e}^*a^\dagger\,e^{-i\mathbf{k}\cdot\mathbf{r}}) \qquad (22)$$

In the phase representation, this Hamiltonian takes on the form $(n \to \infty)$

$$H = \hbar\omega\left(n + \frac{1}{i}\partial_\varphi\right) + c\boldsymbol{\alpha}\cdot\left[\frac{\hbar}{i}\nabla - \gamma(\mathbf{e}\,e^{i(\mathbf{k}\cdot\mathbf{r}-\varphi)} + \text{c.c.})\right] + \beta mc^2 \qquad (23)$$

where

$$\gamma = e(\hbar n/2\omega V)^{1/2} \qquad (24)$$

It is now convenient to change the variables, in order to remove the dependence of the Hamiltonian on the position vector **r**,

$$\varphi \to \varphi - \mathbf{k}\cdot\mathbf{r} \qquad \mathbf{r} \to \mathbf{r} \qquad (25)$$

which results in the following transformations of the derivatives:

$$\partial_\varphi \to \partial_\varphi \qquad \nabla \to \nabla - \mathbf{k}\partial_\varphi \qquad (26)$$

Periodic wave functions will remain periodic after the change of variables. The eigenvalue equation in terms of the new variables reads

$$\left\{\hbar\omega\left[n + (1 - \boldsymbol{\alpha}\cdot\mathbf{n})\frac{1}{i}\partial_\varphi\right]\right.$$

$$\left. + c\boldsymbol{\alpha}\cdot\left[\frac{\hbar}{i}\nabla - \gamma(\mathbf{e}\,e^{-i\varphi} + \text{c.c.})\right] + \beta mc^2 - E\right\}\psi = 0 \qquad (27)$$

where **n** is a unit vector in the direction of **k**. In momentum space, this equation becomes a set of four ordinary differential equations with respect to the phase variable φ. The components of the momentum vector play only the role of additional parameters.

I shall write this set of equations in terms of dimensionless parameters,

$$\left\{(1 - \boldsymbol{\alpha}\cdot\mathbf{n})i\frac{d}{d\varphi} + \epsilon - n - \boldsymbol{\alpha}\cdot[\mathbf{q} - \mathbf{a}(\varphi)] - \beta\kappa\right\}\psi = 0 \qquad (28)$$

where

$$\mathbf{a}(\varphi) = \lambda(\mathbf{e}e^{-i\varphi} + \mathbf{e}^*e^{i\varphi}) \tag{29}$$

$$\epsilon = E/\hbar\omega \qquad \mathbf{q} = \mathbf{p}/\hbar k \qquad \kappa = mc/\hbar k \qquad \lambda = \gamma/\hbar k \tag{30}$$

The solutions of equations (28) can be most easily found when one recognizes the projective character of the matrices $P_\pm$;

$$P_\pm = \tfrac{1}{2}(1 \pm \boldsymbol{\alpha}\cdot\mathbf{n}) \tag{31}$$

I shall introduce two subspaces of the four-dimensional space of ψ's, corresponding to two projectors $P_\pm$,

$$\psi_\pm = P_\pm\psi \tag{32}$$

The projectors act in the following manner on the matrices appearing in equation (28):

$$\begin{aligned}
P_\pm\boldsymbol{\alpha}\cdot\mathbf{q}_L &= \pm P_\pm q_L & P_\pm\boldsymbol{\alpha}\cdot\mathbf{q}_T &= \boldsymbol{\alpha}\cdot\mathbf{q}_T P_\mp \\
P_\pm\boldsymbol{\alpha}\cdot\mathbf{a}(\varphi) &= \boldsymbol{\alpha}\cdot\mathbf{a}(\varphi)P_\mp & P_\pm\beta &= \beta P_\mp
\end{aligned} \tag{33}$$

where $\mathbf{q}_L = q_L\mathbf{n}$ and $\mathbf{q}_T$ are parallel and perpendicular to the $\mathbf{k}$ components of $\mathbf{p}$. Upon multiplying equation (28) by $P_\pm$, with the help of relations (33), we obtain

$$\left(2i\frac{d}{d\varphi} + \epsilon - n - q_L\right)\psi_- = \{\boldsymbol{\alpha}\cdot[\dot{\mathbf{q}}_T - \mathbf{a}(\varphi)] + \beta\kappa\}\psi_+$$

$$(\epsilon - n + q_L)\psi_+ = \{\boldsymbol{\alpha}\cdot[\mathbf{q}_T - \mathbf{a}(\varphi)] + \beta\kappa\}\psi_- \tag{34}$$

The second equation is purely algebraic and may be used to eliminate ψ_+ from the first equation. The resulting differential equation for ψ_- reads

$$\frac{d}{d\varphi}\psi_- = \frac{1}{2i}[q_L - \epsilon + n + (q_L + \epsilon - n)^{-1}\{[\mathbf{q}_T - \mathbf{a}(\varphi)]^2 + \kappa^2\}]\psi_- \tag{35}$$

There are no Dirac matrices in this set of equations, so that the solution can be found for each component separately. The result is

$$\psi_-(\varphi) = \exp\left[\frac{1}{2i}\int_0^\varphi d\varphi\,(q_L - \epsilon + n\right.$$

$$\left. + (q_L + \epsilon - n)^{-1}\{[\mathbf{q}_T - \mathbf{a}(\varphi)]^2 + \kappa^2\})\right]\psi_-(0) \tag{36}$$

The periodicity condition for ψ_- may be easily written down:

$$\int_0^{2\pi} d\varphi \, \{q_L - \epsilon + n + (q_L + \epsilon - n)^{-1}$$

$$\times [\mathbf{q}_T{}^2 - 2\mathbf{q}_T \cdot \mathbf{a}(\varphi) + \mathbf{a}^2(\varphi) + \kappa^2]\} = 4\pi l \qquad (37)$$

where l is an arbitrary integer. After performing all the integrations, we obtain

$$q_L{}^2 - (\epsilon - n)^2 - 2l(q_L + \epsilon - n) + \mathbf{q}_T{}^2 + 2\lambda^2 + \kappa^2 = 0 \qquad (38)$$

The solution for ϵ is

$$\epsilon_{nl} = n - l \pm (l^2 - 2lq_L + \mathbf{q}^2 + 2\lambda^2 + \kappa^2)^{1/2} \qquad (39)$$

Next, I shall transform this result to physical variables. In doing this, we must remember that the true electron momentum $\mathbf{p}_{\text{el}}$ is related to the Fourier variable $\mathbf{p}$ in the following manner [cf. formula (26)]:

$$\mathbf{p}_{\text{el}} = \mathbf{p} - l\hbar\mathbf{k} \qquad (40)$$

$$E_{nL} = (n - l)\hbar\omega \pm c \left(\mathbf{p}^2{}_{\text{el}} + m^2 c^2 + \frac{\epsilon^2 \hbar n}{\omega V} \right)^{1/2} \qquad (41)$$

The most important feature of this formula is the appearance of a correction to the mass. This correction has been found for the first time from a semiclassical calculation by Brown and Kibble[8] and later confirmed in the framework of quantum electrodynamics (cf. a review paper by Eberly[9]). Our derivation of the mass shift is the simplest and most direct. Having found the spectrum, we may insert the value of ϵ into the formula (36) and evaluate the wave functions. These wave functions may be used to calculate transition amplitudes.

Exact solutions can also be found in the presence of an additional uniform magnetic field, pointing in the direction of the photon beam.

References

1. P. A. M. Dirac, *Proc. R. Soc. London Ser. A* **114**, 243 (1927).
2. P. Carruthers and M. Nieto, *Rev. Mod. Phys.* **40**, 411 (1968).
3. H. Paul, *Fortsch. Phys.* **22**, 657 (1974).
4. I. Bialynicki-Birula and Z. Bialynicka-Birula, *Phys. Rev. A* **14**, 1101 (1976).
5. I. Bialynicki-Birula, *Acta Phys. Austriaca Suppl.* **18**, 111 (1977).
6. J. M. Levy-Leblond, *Ann. Phys. (N.Y.)* **101**, 319 (1976).
7. Ya. B. Zeldovich, *Usp. Fiz. Nauk* **110**, 139 (1973).
8. L. S. Brown and T. W. B. Kibble, *Phys. Rev.* **133**, 467 (1964).
9. J. H. Eberly, *Prog. Opt.* **7**, 361 (1969).

10

Radiation Reaction in Nonrelativistic Quantum Theory

D. H. Sharp

1. Introduction

I would like to describe some calculations[1,2] which E. J. Moniz and I carried out for the purpose of understanding more clearly the relationship between classical and quantum electrodynamics, particularly in regard to their treatment of radiation reaction. Let us begin by reviewing some of the questions of interest here.

According to the classical theory of radiation reaction due to Abraham, Lorentz, and Dirac,[3-5] a nonrelativistic point electron, interacting with its self-field and subject to an external force $\mathbf{F}(t)$, obeys the equation of motion

$$m_0\ddot{\mathbf{R}} = \mathbf{F}(t) - \delta m\ddot{\mathbf{R}}(t) + (2e^2/3c^3)\dddot{\mathbf{R}}(t) \qquad (1.1)$$

where δm is the electron's electrostatic self-energy.

This theory of radiation reaction suffers from a number of defects besides the fact that $\delta m = \infty$ for a point electron, a fact that can after all be swept under the rug, according to the philosophy of renormalization, by working with the experimental mass.

The first defect is that equation (1.1) admits runaway solutions, i.e., solutions for which the acceleration of a particle increases exponentially, even in the absence of external forces.

The second defect is that the solutions violate causality. This comes

D. H. Sharp • Theoretical Division, Los Alamos Scientific Laboratory, University of California, Los Alamos, New Mexico 87545. Work supported by the U.S. Department of Energy.

about when runaways are eliminated from the theory by imposition of a suitable asymptotic condition. To see this, notice that the general solution to equation (1.1) is

$$\ddot{\mathbf{R}}(t) = e^{t/\tau}\left[\dddot{\mathbf{R}}(0) - \frac{1}{\tau m}\int_0^t dt' \, e^{-t'/\tau}\mathbf{F}(t')\right] \tag{1.2}$$

with $\tau = 2e^2/3mc^3$. If $\dddot{\mathbf{R}}(0)$ is chosen arbitrarily, equation (1.2) gives an acceleration which grows asymptotically like $e^{t/\tau}$ even if the force acts only for a finite period of time. This behavior can be avoided if you impose the condition

$$\dddot{\mathbf{R}}(0) = \frac{1}{\tau m}\lim_{t\to\infty}\int_0^t dt' \, e^{-t'/\tau}\mathbf{F}(t') \qquad (\text{i.e.,}\ \lim_{t\to\infty}\ddot{\mathbf{R}}(t)\to 0)$$

But then one can write the solution (1.2) in the form

$$\ddot{\mathbf{R}}(t) = \frac{1}{\tau m}\int_t^\infty dt' \, e^{-(t'-t)/\tau}\mathbf{F}(t')$$

or introducing $s = (t' - t)/\tau$, as

$$\ddot{\mathbf{R}}(t) = \frac{1}{m}\int_0^\infty ds \, e^{-s}\mathbf{F}(t + \tau s) \tag{1.3}$$

This form of the solution displays clearly the acausal behavior known as preacceleration: the electron accelerates before the force acts.

While these defects mar the internal consistency of classical electro-dynamics, the point can be made that equation (1.3) does in fact correctly describe classical radiation damping, in so far as it has been tested, and the view is often adopted that, since preacceleration occurs on such a short time scale ($\sim 10^{-23}$ second for an electron), the acausal effects would occur in the quantum domain, which is where one has to look for a resolution of the problem.

It is a very reasonable proposition that runaway solutions should not occur in quantum theory. One would not expect a Heisenberg-picture operator to display an exponentially growing dependence on time, since its time development is given by

$$\mathcal{O}(t) = e^{iHt}\mathcal{O}(0)e^{-iHt}$$

with e^{iHt} unitary. Nevertheless, to date no rigorous proof of the absence of runaways in quantum electrodynamics has been given.

Unfortunately, such a rigorous proof is not the subject of the present paper either. Instead, I will describe some rather straightforward calculations which appear to shed some light on the following questions:

(i) What is the mechanism by which runaway solutions are eliminated in quantum mechanics?

(ii) How does quantum theory manage to suppress the runaways and at the same time give equation (1.1) in the correspondence limit?

(iii) What sort of formula do you get for the electrostatic self-energy in quantum theory?

(iv) What about preacceleration?

In other words, we will discuss whether quantum theory resolves any of the problems of consistency which appear to be present already in classical electrodynamics. We will not be addressing the more fundamental questions of the possible finiteness, and overall consistency, of quantum electrodynamics.

2. Classical Electrodynamics of Extended Charges

It will turn out that some aspects of our results on the quantum theory of radiation reaction can be best understood by comparing them to the classical results for the motion of an *extended* charge. For a spherically symmetric static charge distribution $\rho(\mathbf{x}, t) = \rho[\mathbf{x} - \mathbf{R}(t)]$, where $\mathbf{R}(t)$ is the coordinate of the mean position of the charge, equation (1.1) is replaced by[6]

$$m_0 \ddot{\mathbf{R}}(t) = \mathbf{F}(t) - \frac{2e^2}{3c^2} \sum_{n=0}^{\infty} \frac{(-1)^n}{n! c^n}$$

$$\times \gamma_n \frac{d^{n+2}\mathbf{R}(t)}{dt^{n+2}} + \text{(nonlinear terms)} \qquad (2.1)$$

where

$$\gamma_n = \int \int d\mathbf{x} \, d\mathbf{x}' \rho(\mathbf{x}, t) |\mathbf{x} - \mathbf{x}'|^{n-1} \rho(\mathbf{x}', t) \propto L^{n-1}$$

and L is the effective charge radius.

We have shown explicitly in equation (2.1) only terms which are linear in the particle's velocity or its time derivatives. These terms all arise from the electric self-field. The nonlinear terms, which arise both from the electric and magnetic self-fields, are all of order $|\dot{\mathbf{R}}/c|^2$ times the linear terms. These are neglected in this discussion, since we are considering the motion of a nonrelativistic electron.

For simple charge distributions, the coefficients γ_n can be explicitly evaluated and the series summed. Thus, for a spherical shell, one obtains

$$\gamma_n = (2e^2)(2L)^{n-1}/(n + 1) \tag{2.2}$$

and the equation of motion can be written in the form[2,7,8]

$$\ddot{\mathbf{R}}(t) = \mathbf{F}(t)/m(1 - c\tau/L) + \xi[\dot{\mathbf{R}}(t - 2L/c) - \dot{\mathbf{R}}(t)] \tag{2.3}$$

neglecting nonlinear terms, where

$$\xi = \frac{(c/2L)(c\tau/L)}{(1 - c\tau/L)} \qquad \tau = \frac{2e^2}{3mc^3} \qquad m = m_0 + \frac{2e^2}{3Lc^2}$$

The solutions to equation (2.3) have been analyzed fully.[2,8] One finds that if $L > c\tau$ ($\xi > 0$), equation (2.3) has no runaway nor preaccelerating solutions, while if $L \ll c\tau$, equation (2.3) reduces to equation (1.1) with $m_0 + \delta m = m$. Runaway and acausal solutions of equation (2.3) occur if $L < c\tau$.

The pertinence of these results to the quantum-mechanical case is the following. We will find that the structure of the radiation reaction problem for a quantum mechanical *point* electron, i.e., an electron with zero charge radius, is similar to that of a classical *extended* charge. Specifically, the quantum-mechanical equation of motion for a point charge has the general form of equation (2.1), with the electron Compton wavelength λ formally playing the role of the charge radius L. It is the fact that there is a new length scale in the quantum theory which allows this to happen.

3. Quantum Theory of Radiation Reaction

Our plan is to first derive the Heisenberg-picture operator equation of motion for a nonrelativistic electron, including the self-force terms, and then to analyze some properties of its solutions. In other words, we want to study the quantum counterparts of equation (1.1) or equation (2.1).

3.1. Equation of Motion

To derive the equation of motion, we follow the Abraham–Lorentz procedure for deriving the self-force on an electron, except that we must remember to take proper account of the fact that we are working with operators. For the purposes of the present discussion, this just means that we pay attention to the order of noncommuting quantities.

Our starting point for this calculation is the Hamiltonian[9]

$$H = \frac{1}{2m_0} [\mathbf{P} - \frac{e}{c} \mathbf{A}(\mathbf{R})]^2 + \frac{1}{8\pi} \int d\mathbf{r} \, \{\mathbf{E}^2(\mathbf{r}, t) + [\nabla \times \mathbf{A}(\mathbf{r}, t)]^2\} \quad (3.1)$$

where

$$\mathbf{A}(\mathbf{R}) = \int d\mathbf{r} \, \rho[\mathbf{r} - \mathbf{R}(t)]\mathbf{A}(\mathbf{r}, t)$$

and

$$\mathbf{E} = \mathbf{E}_{\text{long}} + \mathbf{E}_{\text{trans}}$$

This describes a nonrelativistic charged particle of mechanical mass m_0 and (spherically symmetric) charge distribution[10] [defined so that $\int d\mathbf{r} \, \rho(\mathbf{r} - \mathbf{R}) = 1$] interacting with an electromagnetic field, computed in Coulomb gauge. $\mathbf{P}(t)$ and $\mathbf{R}(t)$ are, respectively, the Heisenberg-picture momentum and position operators of the particle.

Proceeding in standard fashion, we use equation (3.1) to derive the Heisenberg equations of motion and arrive at the operator form of the Lorentz force equation

$$\frac{d}{dt}(m_0\dot{\mathbf{R}}) = e\mathbf{E}(\mathbf{R}) + \frac{e}{2c}[\dot{\mathbf{R}} \times \mathbf{B}(\mathbf{R}) - \mathbf{B}(\mathbf{R}) \times \dot{\mathbf{R}}] \quad (3.2)$$

and the usual operator field equations for the electromagnetic potentials $\mathbf{A}$ and ϕ (here written in Lorentz gauge[11]),

$$-\Box\mathbf{A}(\mathbf{r}, t) = 4\pi e\mathbf{j}(\mathbf{r}, t) \quad (3.3a)$$

$$-\Box\phi(\mathbf{r}, t) = 4\pi e\rho(\mathbf{r}, t) \quad (3.3b)$$

In writing these equations, we have used the following notation:

$$\mathbf{B}(\mathbf{R}) = \nabla \times \mathbf{A}(\mathbf{R}), \qquad \mathbf{E}(\mathbf{R}) = -\nabla\phi(\mathbf{R}) - \frac{1}{c}\frac{\partial\mathbf{A}(\mathbf{R})}{\partial t}$$

and

$$\mathbf{j}(\mathbf{r}, t) = \tfrac{1}{2}[\rho(\mathbf{r} - \mathbf{R}(t)), \dot{\mathbf{R}}(t)]_+ = \text{the single-particle current}$$
$$\text{density operator}$$

As in the classical case, our goal now is to use the field equations (3.3) to eliminate the self-fields from the Lorentz force equation. To do this, we first observe that the exact solution to equation (3.3), satisfying retarded boundary conditions, may be written

$$\mathbf{A}(\mathbf{r}, t) = \mathbf{A}_{\text{in}}(\mathbf{r}, t) + \frac{e}{c} \int d\mathbf{r}' \, \frac{\mathbf{j}(\mathbf{r}', t'_{\text{ret}})}{|\mathbf{r} - \mathbf{r}'|} \quad (3.4a)$$

$$\equiv \mathbf{A}_{\text{in}} + \mathbf{A}_{\text{self}}$$

and

$$\phi(\mathbf{r}, t) = \phi_{\text{in}}(\mathbf{r}, t) + e \int d\mathbf{r}' \frac{\rho(\mathbf{r}', t'_{\text{ret}})}{|\mathbf{r} - \mathbf{r}'|} \tag{3.4b}$$

$$\equiv \phi_{\text{in}} + \phi_{\text{self}}$$

where the retarded time is given by $t'_{\text{ret}} = t - (|\mathbf{r} - \mathbf{r}'|/c)$. Notice that $\lim_{t \to -\infty} \mathbf{A} \to \mathbf{A}_{\text{in}}$, which is a free field.

We can relate an operator evaluated at the retarded time t'_{ret} to its value at time t by the formula

$$\mathcal{O}(t'_{\text{ret}}) = \exp[+iH(t'_{\text{ret}} - t)]\mathcal{O}(t)\exp[-iH(t'_{\text{ret}} - t)]$$

$$= \sum_{n=0}^{\infty} \frac{(-i)^n}{n!} \frac{|\mathbf{r} - \mathbf{r}'|^n}{c^n} (\text{ad}^n H)\mathcal{O}(t) \tag{3.5}$$

where

$$(\text{ad } H)\mathcal{O} = [H, \mathcal{O}]_- \quad \text{and} \quad (\text{ad}^2 H)\mathcal{O} = [H, [H, \mathcal{O}]_-]_-$$

Next, we use the expansion (3.5) and equation (3.4) to evaluate the electric and magnetic self-fields occurring in the Lorentz force equation. The resulting form of the quantum-mechanical equation of motion is

$$m_0 \ddot{\mathbf{R}}(t) = e\mathbf{E}_{\text{in}} + \frac{e}{2c}[\dot{\mathbf{R}} \times \mathbf{B}_{\text{in}} - \mathbf{B}_{\text{in}} \times \dot{\mathbf{R}}] + \frac{2e^2}{3c^2} \sum_{n=0}^{\infty} \frac{(-i)^{n+1}}{n!c^n}$$

$$\times \iint d\mathbf{r}\, d\mathbf{r}' \tfrac{1}{2}[\rho(\mathbf{r} - \mathbf{R}(t))|\mathbf{r} - \mathbf{r}'|^{n-1}, (\text{ad}^{n+1} H)\mathbf{j}(\mathbf{r}', t)]_+ \tag{3.6}$$

In writing equation (3.6), we have dropped the contributions from the magnetic self-field. These have been evaluated and found to be formally of order $\dot{\mathbf{R}}^2/c^2$ times the leading contributions from the electric self-field, as one would expect. Such contributions should therefore be negligible for a slowly moving electron. The same should be true of the nonlinear terms associated with the electric self-field. The statement that these terms are negligible means, in the quantum mechanical context, that there is a subset of states in Hilbert space for which the matrix elements $\langle m | \dot{\mathbf{R}}^2/c^2 | n \rangle$ are ''small'' and that one can work consistently to a given level of accuracy within this set of states. It is a crucial assumption of this calculation that such a set of states exists.

Equation (3.6) is the starting point for our study of radiation reaction in quantum mechanics. The main labor in this calculation consists in evaluating the nested commutators in equation (3.6), so that it can be put in a useful form. The details can be found in Reference 2, and I won't reproduce them here. However, it might be instructive to see what the first few terms in the series (3.6) look like.

After evaluation of $[H, \mathbf{j}(\mathbf{r}, t)]$, the $n = 0$ term in equation (3.6) gives[12]

$$(m_0\ddot{\mathbf{R}})_{n=0} = -(2e^2/3c^2)\langle|\mathbf{r} - \mathbf{r}'|^{-1}\rangle\ddot{\mathbf{R}}$$

with the notation $\langle|\mathbf{r} - \mathbf{r}'|\rangle = \iint d\mathbf{r}\,d\mathbf{r}'\,\rho(\mathbf{r})|\mathbf{r} - \mathbf{r}'|\rho(\mathbf{r}')$. This is the same as the classical result. Similarly, the $n = 1$ term gives

$$(m_0\ddot{\mathbf{R}})_{n=1} = (2e^2/3c^3)\langle 1\rangle\dddot{\mathbf{R}}$$

which is again the same as in the classical case. The first new quantum terms come in when $n = 2$, where you find

$$(m_0\ddot{\mathbf{R}})_{n=2} = \underbrace{-(1/3c^4)\langle|\mathbf{r} - \mathbf{r}'|\rangle\ddddot{\mathbf{R}}}_{\substack{\text{classical}\\\text{result}}} \underbrace{-(1/9c^4)\langle|\mathbf{r} - \mathbf{r}'|^{-1}\rangle[3\{\dot{\mathbf{R}}^2,\ \dddot{\mathbf{R}}\} + \{\ddot{\mathbf{R}}\cdot\dot{\mathbf{R}},\ \ddot{\mathbf{R}}\} + \{\dot{\mathbf{R}}\cdot\ddot{\mathbf{R}},\ \ddot{\mathbf{R}}\} + \dot{\mathbf{R}}\cdot\ddot{\mathbf{R}}\,\ddot{\mathbf{R}} + \dot{\mathbf{R}}\,\ddot{\mathbf{R}}\cdot\dot{\mathbf{R}}]}_{\substack{\text{neglecting order of operators, exactly}\\ \text{the } \dot{\mathbf{R}}^2/c^2 \text{ correction arising from } \mathbf{E}_{\text{self}}\\ \text{to the classical result for } \dddot{\mathbf{R}}}}$$

$$\underbrace{-(8\pi/3c^2)(\hbar/m_0c)^2[\int d\mathbf{r}\,\rho^2(\mathbf{r})]\dddot{\mathbf{R}}}_{\text{new quantum term}}$$

Let L denote the particle's charge radius and let $\lambda = (\hbar/m_0c)$ be the electron's Compton wavelength. We see that the new term diverges as $(1/L^3)$ as $L \to 0$ and is $\sim(\lambda/L)^2$ times the classical self-energy term. This suggests a small quantum-mechanical correction to the self-energy if $\lambda \ll L$, but if $\lambda \gg L$ (point-charge limit), one will have to sum the series.

This is how one proceeds, carrying out the term by term evaluation of equation (3.6), until one has inferred the combinatorics governing the general term. The result is[1,2] (dropping all terms of order $\dot{\mathbf{R}}^2/c^2$ or smaller compared to the leading terms)

$$m_0\dddot{\mathbf{R}}(t) = e\mathbf{E}_{\text{in}} + \frac{e}{2c}[\dot{\mathbf{R}} \times \mathbf{B}_{\text{in}} - \mathbf{B}_{\text{in}} \times \dot{\mathbf{R}}]$$

$$-\frac{2e^2}{3c^2}\sum_{n=0}^{\infty}\frac{(-1)^n}{n!c^n}A_n\overset{(n+2)}{\mathbf{R}}(t) \tag{3.7}$$

with

$$A_n = \left[1 + \frac{\lambda}{3(n+2)}\frac{\partial}{\partial\lambda}\right]B_n$$

$$B_n = \sum_{k=0}^{\infty}\frac{n!}{(n+2k)!}\binom{n+1+2k}{2k}\left(\frac{-\lambda^2}{4}\right)^k$$

$$\times \iint d\mathbf{r}\,d\mathbf{r}'\,\rho(r)|\mathbf{r} - \mathbf{r}'|^{n-1+2k}(\nabla_r^2)^{2k}\rho(\mathbf{r}')$$

We have used the notation that $\overset{(m)}{\mathbf{R}}(t) = d^m\mathbf{R}(t)/dt^m$ and, again, that $\lambda = \hbar/m_0c$. Note that each structure coefficient is a power series in (λ^2/L^2) and that if we retain only the $k = 0$ term in the series for B_n, which corresponds to taking the $\hbar \to 0$ limit of he expression, we recover the equation of motion for a classical extended charge.

3.2. Evaluation of the Structure Coefficients

3.2.1. Electrostatic Self-Energy

The electrostatic self-energy, defined as the coefficient of the acceleration arising from the self-force, is given by the $n = 0$ term in the series in equation (3.7). Specifically, one obtains

$$\delta m = \frac{2e^2}{3c^2} A_0 = \frac{2e^2}{3c^2}\left(1 + \frac{\lambda}{6}\frac{\partial}{\partial\lambda}\right)\left(1 + \lambda\frac{\partial}{\partial\lambda}\right)\Omega_0 \qquad (3.8)$$

where

$$\Omega_0 = \int \frac{d\mathbf{k}}{(2\pi)^3}\,\tilde{\rho}(k)^2 \sum_{l=0}^{\infty} \frac{(-1)^l}{(2l)!}\left(\frac{\lambda k^2}{2}\right)^{2l} \int \frac{d\mathbf{r}}{r}\,e^{i\mathbf{k}\cdot\mathbf{r}}r^{2l}$$

and with $\tilde{\rho}(k)$ the Fourier transform of $\rho(r)$. Doing the integral over r and summing the series leads to

$$\Omega_0 = (2/\pi)\mathscr{P}\int_0^\infty dk\,\frac{\tilde{\rho}(k)^2}{1 - \lambda^2k^2/4} \qquad (3.9)$$

where the improper k-integral has been regularized by taking the Cauchy principal value.

This formula for the electrostatic self-energy has a number of remarkable features:

(*i*) If one lets $\lambda \to 0$ in equations (3.8) and (3.9) and *then* goes to the point-charge limit, one obtains the classic divergent expression for δm. However, if one first takes the point-charge limit $[\tilde{\rho}(k) = 1]$ in equation (3.9), *keeping λ fixed,* one finds $\delta m = 0$. Thus, according to this calculation, one finds that the electrostatic self-energy of a point charge is zero in nonrelativistic quantum electrodynamics. This is a surprising result. To keep things in perspective we emphasize that it is *not* claimed that this calculation shows that the electron's self-mass is finite. There are contributions to this self-mass other than the one treated here, and these other contributions may well be infinite. Nevertheless, it would be most interesting to see what an essentially nonperturbative calculation, as this is, would tell us about these contributions.

(*ii*) It is also interesting to study the self-energy in the case when the particle has a convergent form factor, such as the Yukawa form factor $\bar{\rho}(k) = (1 + k^2 L^2)^{-1}$. ($L$ is the effective charge radius.) Using equations (3.8) and (3.9), one finds[2] that the maximum value for δm occurs when $L \sim \lambda$, and that then

$$(\delta m)_{\text{MAX}} \sim \alpha m_0 \tag{3.10}$$

where α is the fine-structure constant. This seems to be a physically reasonable result. For example, it would lead to hadronic electromagnetic mass shifts on the order of a few MeV if m_0 were chosen to be a few hundred MeV. Furthermore, equation (3.10) excludes the possibility of a purely electromagnetic origin for the electron's mass within the framework of nonrelativistic quantum electrodynamics.

(*iii*) For $0 < L \ll \lambda$, the electrostatic self-energy can actually become negative.[2]

(*iv*) Additional insight into these results can be obtained by transforming Ω_0, equation (3.9), back into coordinate space. One finds that

$$\Omega_0 = \iint d\mathbf{r}\, d\mathbf{r}' \, \rho(r) |\mathbf{r} - \mathbf{r}'|^{-1} \left[1 + \frac{\lambda^2}{4} \nabla_{r'}^2 \right]^{-1} \rho(r') \tag{3.11}$$

The integral operator in this expression is defined by

$$(1 + \lambda^2 \nabla^2/4)^{-1} \rho(x) = \int dy\, S_\lambda(x - y)\rho(y) \equiv \rho_{\text{eff}}(x)$$

with

$$S_\lambda(r) = \mathscr{P} \int \frac{d\mathbf{k}}{(2\pi)^3} \frac{e^{i\mathbf{k}\cdot\mathbf{r}}}{1 - \lambda^2 k^2/r}$$

$$= \frac{-\cos(2r/\lambda)}{\pi\lambda^2 r} \tag{3.12}$$

Thus we see that in this calculation all of the physics involved in the interaction of the charged particle with its quantized self-field is summarized in the "spreading function" $S_\lambda(r)$ which generates an effective charge distribution $\rho_{\text{eff}}(r)$ which is smeared out over a Compton wavelength. In this respect, our results are quite similar to those obtained many years ago by Weiskopf.[13] In his work, however, the spreading out of the charge distribution was caused by virtual electron-positron pairs, whereas in our strictly nonrelativistic treatment there are, of course, no positrons.

Finally, we note that the charge distribution ρ_{eff} generates an effective scalar potential

$$\phi_{\text{eff}}(r) = \int d\mathbf{r}\, |\mathbf{r} - \mathbf{r}'|^{-1} \rho_{\text{eff}}(r')$$

from which the electrostatic self-energy can be calculated as

$$\Omega_0 = \int d\mathbf{r} \, \rho(r)\phi_{\text{eff}}(r)$$

3.2.2. The Remaining Coefficients

I will simply assert in passing that the other structure coefficients in equation (3.7) have been evaluated,[1,2] and that in the point-charge limit one finds

$$A_n = \begin{cases} (-1)^{(n-1)/2} \dfrac{2n(4n+5)}{3(n+1)(n+2)} (2n-1)!!\lambda^{n-1} & n \text{ odd} \\ 0 & n \text{ even} \end{cases} \tag{3.13}$$

Thus the equation of motion (3.7) is indeed similar in structure to that of a classical extended charged particle, equation (2.1), with the Compton wavelength λ playing the role of a size parameter.

3.3. Solutions of the Equation of Motion

3.3.1. Motion in the Absence of External Forces

Now I want to discuss some properties of the solutions to equation (3.7). First I will discuss the motion of a "free" electron, i.e., one that experiences self-interactions, but is not acted upon by any external force. This is the situation in which one encounters runaway solutions classically, and we want to see what happens in the quantum-mechanical case.

To investigate this question, we take matrix elements of the equation of motion between the exact stationary states of the Hamiltonian (3.1). We assume that among these states there are ones for which the matrix elements of the in-fields are negligible, and we confine our attention to these states. This is possible owing to the linearity of equation (3.7).

While we do not know how to construct the exact stationary states of the Hamiltonian (3.1), we do know that for such states one can write

$$\langle m|\dot{\mathbf{R}}(t)|n\rangle = \exp(iE_{mn}t/\hbar)\langle m|\dot{\mathbf{R}}(0)|n\rangle$$

with $E_{mn} = E_m - E_n$. We see that *if* there were runaway solutions to equation (3.7), there would have to be states such that $\langle m|\dot{\mathbf{R}}(0)|n\rangle \neq 0$ and for which $\beta \equiv iE_{mn}/\hbar$ has a positive real part.

Supposing that $\langle m\dot{\mathbf{R}}(0)|n\rangle \neq 0$, and taking the indicated matrix elements of equation (3.7), results in a power series in the variable $\eta =$

$\beta\lambda/c$:

$$1 = \tfrac{2}{3}\alpha\eta \sum_{\substack{n=1 \\ \text{odd}}}^{\infty} (-1)^{(n-1)/2} \frac{(2n-1)!!}{n!} \left[\frac{1}{3}\left(\frac{2n}{n+1} \right)\left(\frac{4n+5}{n+2} \right) \right] \eta^{n-1}$$

$$\equiv \tfrac{2}{3}\,\alpha f(\eta) \tag{3.14}$$

In writing equation (3.14), I have factored out the root $\beta = 0$. This corresponds to motion at constant velocity and is the expected result for a free electron. The question is whether there are other solutions to equation (3.14), corresponding to runaways or other unphysical motions.

The series (3.14) converges for $|\eta| < \tfrac{1}{2}$. Inside its radius of convergence, the series can be summed and one obtains

$$f(\eta) = -(4i/3)[(1 - 2i\eta)^{-1/2} - (1 + 2i\eta)^{-1/2}]$$
$$-(7/3\eta)[(1 - 2i\eta)^{1/2} + (1 + 2i\eta)^{1/2} - \tfrac{2}{7}]$$
$$+ (2i/3\eta^2)[(1 - 2i\eta)^{3/2} - (1 + 2i\eta)^{3/2}] \tag{3.15}$$

Thus, to examine the question of runaways, one must determine the roots of the equation

$$1 = \tfrac{2}{3}\alpha f(\eta) \tag{3.16}$$

where α is the fine-structure constant and $f(\eta)$ is given by equation (3.15).

Here is what you find[1,2]: (*i*) For *physical* values of the fine-structure constant, in fact for all $\alpha \lesssim 1$, equation (3.16) has *no* roots inside the radius of convergence $|\eta| = \tfrac{1}{2}$. (*ii*) For large α, interpreted either as a strong-coupling limit or a semiclassical limit, one does obtain a real root of equation (3.16) for $|\eta| < \tfrac{1}{2}$. In fact, what one finds in this case is a small $\hbar$ expansion about the classical runaway solution

$$\beta \sim (1/\tau)[1 + (\text{numerical coefficient})\, \hbar^2 + \cdots]$$
$$= (1/\tau)[1 + (\text{numerical coefficient})'\, (1/\alpha)^2 + \cdots] \tag{3.17}$$

where $\tau = (2e^2/3mc^3)$. (*iii*) The large and small α regimes are separated in that there is a critical value of α, $\alpha_{\text{crit}} > 1$, such that

$$\left. \frac{d\eta}{d\alpha} \right|_{\alpha \to \alpha_{\text{crit}}} \to \infty$$

This behavior is like that of a first-order phase transition, and it means that the radius of convergence of equation (3.17) cannot include the physical value of α.

Thus the analysis of the roots of equation (3.16) shows that runaway solutions are not present in nonrelativistic quantum electrodynamics for physical values of the fine-structure constant. Our understanding of this result is that the interaction of a charged point particle with its quantized self-field generates an effective charge distribution spread out over a Compton wavelength [equation (3.12)]. This structure is reflected in the particle's equation of motion, resulting in a form for the quantum-mechanical equation of motion for a point charge [equations (3.7) and (3.13)] which is similar to that of an extended classical charge [equation (2.1)]. Several analyses[2,7,8] indicate that a classical charged particle of sufficient size (charge radius > classical electron radius) does not exhibit runaway behavior.

It is also interesting to inquire about the significance of the condition $|\eta| < \frac{1}{2}$. Recalling the various definitions, we see that it says that

$$E_{mm} < \tfrac{1}{2} mc^2 \tag{3.18}$$

This condition represents a restriction on the energy eigenstates of particle plus field, between which one can consistently evaluate matrix elements of equation (3.17), and expresses the fact that our results are limited to the nonrelativistic domain. It is remarkable that the criterion (3.18) is generated by the dynamical equations themselves.

3.3.2. External Forces

Now let us consider how the electron moves in response to a time-dependent external force $\mathbf{F}(t)$. If we again neglect the in-fields, the equation of motion can be solved by Fourier transformation and the solution can be written in the form

$$m_0 \ddot{\mathbf{R}}(t) = \int_{-\infty}^{\infty} dt' \; G(t - t')\mathbf{F}(t') \tag{3.19}$$

were the response function $G(t - t')$ is given by[1,2]

$$G(t - t') = \frac{1}{2\pi} \int_{-c/2\lambda}^{c/2\lambda} d\omega \; \frac{\exp[i\omega(t - t')]}{1 - \tfrac{2}{3}\alpha f(i\omega\lambda/c)} \tag{3.20}$$

In equation (3.20), the function $f(i\omega\lambda/c)$ is defined by equation (3.15).

To derive equations (3.19) and (3.20), we have had to require that $\mathbf{F}(\omega)$ vanish for $|\omega| > c/2\lambda$. This is because the series defining $f(i\omega\lambda/c)$ does not converge unless $|\omega| < c/2\lambda$. This condition is of course closely related to that derived above, $E_{mn} < \tfrac{1}{2}mc^2$, as a requirement of nonrelativistic motion, and it also follows directly from the condition that the applied force change by a small amount in the time required for light to cross a Compton wavelength.

It is not difficult to see that the response function (3.20) does not allow for observable preacceleration if $\alpha \ll 1$. An approximate evaluation of (3.20) shows that the quantum response function is spread about the origin ($t = t'$) with a minimum width given by the characteristic time $\Delta T \sim 2\lambda/c$. This time is determined jointly by the uncertainty principle and the equation of motion (through the condition $\omega < c/2\lambda$), and the time at which the particle starts to move in response to the applied force cannot be determined more accurately than ΔT. The time scale for preacceleration, on the other hand, is

$$\tau \sim (e^2/mc^3) = \alpha(\lambda/c) \ll \Delta T \qquad \text{if } \alpha \ll 1$$

so there can be no observable violation of causality. Note, however, that this conclusion would not follow if $\alpha \sim 1$.

If the force is cut off at a frequency small compared to c/λ, the correspondence limit of equation (3.19) can be obtained by expanding the denominator of the response function (3.20). One finds that

$$m_0 \ddot{\mathbf{R}}(t) = \mathbf{F}(t) + \tau\dot{\mathbf{F}}(t) + \cdots$$

$$= \int_0^\infty ds \, e^{-s}\mathbf{F}(t + s\tau) \qquad (3.21)$$

Thus, *in the classical regime,* equations (3.19) and (3.20) give the same result as equation (1.3)—the solution to the classical Lorentz–Dirac equation which results when the runaway solution is eliminated by fiat. The interesting point is that equations (3.19) and (3.20) place a limit, originating from quantum theory, on the applicability of the classical solution (1.3).

4. Discussion of the Correspondence between Quantum and Classical Electrodynamics

We have studied the quantum theory of a nonrelativistic charged particle coupled to the quantized electromagnetic field. This model, which we have been calling "nonrelativistic quantum electrodynamics," is defined by the Hamiltonian (3.1).

We have mentioned that nonrelativistic classical electrodynamics appears to be internally consistent in describing the motion of extended charged particles. However, taking the point-charge limit of the theory of a classical extended charge results in a set of equations whose solutions display runaway behavior and preacceleration. In quantum mechanics, on the other hand, the point-charge theory is consistent, displaying neither runaway behavior nor observable acausality. Furthermore, the correspondence limit of the *solutions* of the quantum-mechanical equation of motion reproduce those properties, and only those properties, of the solutions of the classical Lorentz-Dirac equation which are physically reasonable. Thus, a consistent picture of a classical point electron emerges as the correspondence limit of a quantum-mechanical point electron, but not as the point limit of a classical extended charge.[14]

5. Conclusion

In conclusion, I want to mention some questions which I believe this work raises. The model studied here, based on the Hamiltonian (3.1), lacks relativistic invariance. Consequently, no overall consistency for this model can be claimed. Moreover, we do not know which of our conclusions about the consistency of nonrelativistic quantum electrodynamics would continue to hold in relativistic quantum electrodynamics. Thus, an attempt to extend our calculations to the domain of fully relativistic field theory is clearly important.

Another point concerns the interesting fact that, according to our calculation, nonrelativistic quantum electrodynamics apparently would display runaway behavior and preacceleration if the fine-structure constant were greater than about one. This suggests an upper bound on α. Is this bound real, or would it disappear if more physics, such as pair creation, were included in the model? If the bound is real, what general property of the theory is being revealed?

Finally, as has recently been stressed by Grotch and Kazes,[15] it would be quite instructive to understand more clearly the relationship between our method of calculation and standard perturbation theory. Some progress has been made on this question, and in fact it turns out that quite different assumptions and approximations underlie the two methods. There is, therefore, no particular reason why the results obtained from these two methods should agree in any given order of approximation. I will not go into detail about this work here, since it is discussed elsewhere.[16,17]

References and Notes

1. E. J. Moniz and D. H. Sharp, *Phys. Rev. D* **10**, 1133 (1974).
2. E. J. Moniz and D. H. Sharp, *Phys. Rev. D* **15**, 2850 (1977).
3. H. A. Lorentz, *Theory of Electrons*, 2nd ed., Dover, New York (1952).
4. P. A. M. Dirac, *Proc. R. Soc. London Ser. A* **167**, 148 (1938).
5. Excellent expositions of the Abraham–Lorentz–Dirac theory can be found in T. Erber, *Fortschr. Phys.* **9**, 343 (1961); J. D. Jackson, *Classical Electrodynamics*, Wiley, New

York (1962); F. Rohrlich, *Classical Charged Particles–Foundations of Their Theory*, Addison-Wesley, Reading, Massachusetts (1965).
6. See the textbook of J. D. Jackson, Reference 5, for details.
7. The spherical shell has recently been discussed in References 2 and 8. For references to previous work on the motion of classical extended charges, see Reference 2 and the review article of T. Erber, Reference 5.
8. H. Levine, E. J. Moniz, and D. H. Sharp, *Am. J. Phys.* **45**, 75 (1977).
9. W. Pauli and M. Fierz, *Nuovo Cimento* **15**, 157 (1938).
10. In obtaining the equation of motion we introduce a static charge distribution ρ into the Hamiltonian. We will pass to the point limit at a later stage of the calculation.
11. We have checked that our calculation leads to the same results if one works in Coulomb gauge.
12. The only properties of the Hamiltonian that are used to carry out the evaluation are that $\partial \mathbf{j}(\mathbf{r}, t)/\partial t = i[H, \mathbf{j}(\mathbf{r}, t)]$ and that $\partial \rho/\partial t = i[H, \rho] = -\nabla \cdot \mathbf{j}$, with the current density given by $\mathbf{j}(\mathbf{r}, t) = \frac{1}{2}[\rho(\mathbf{r} - \mathbf{R}(t)), \dot{\mathbf{R}}(t)]_+$.
13. V. F. Weiskopf, *Phys. Rev.* **56**, 72 (1939).
14. This point has also been emphasized recently by F. Rohrlich, *Acta Phys. Austriaca* **41**, 375 (1975).
15. H. Grotch and E. Kazes, *Phys. Rev. D* **16**, 3605 (1977).
16. H. Grotch, E. Kazes, F. Rohrlich, and D. H. Sharp, in preparation.
17. F. Rohrlich, Chapter 13 of this volume.

11

Heisenberg Equation of Motion Calculation of the Electron Self-Mass in Nonrelativistic Quantum Electrodynamics

H. Grotch and E. Kazes

The calculation of the self-mass of a point electron has been a theoretical problem which many generations of physicists have worked on intensively. Perturbative analyses, carried out either in classical or quantum electrodynamics, nonrelativistically or relativistically, have invariably led to an infinite result. Several years ago, a new approach was given by Moniz and Sharp.[1] Their work was nonperturbative in the sense that the traditional expansion in the fine-structure constant was not utilized, but rather other approximations, which we shall elaborate on later, were used. Their approach was based on the Heisenberg equations of motion. An important conclusion of their studies was that the quantum-mechanical mass shift of the nonrelativistic point electron is zero. The classical mass shift was also shown to vanish.

These surprising results, which differ from standard perturbative calculations, stimulated our interest in this problem. For the case of a finite-size electron Moniz and Sharp obtained a nonvanishing mass shift proportional to α, the fine-structure constant. This result, however, differs from the standard quantum electrodynamic (QED) result obtained by a conventional second-order perturbation theory calculation of the energy shift of an electron. In an attempt to resolve this discrepancy and to further understand the nature of the Moniz and Sharp result we calculated the electron mass shift using second-order perturbation theory in

H. Grotch and E. Kazes • Department of Physics, Pennsylvania State University, University Park, Pennsylvania 16802

the context of the Heisenberg equation of motion. Our result, which we discuss below, agrees with the conventional QED result and disagrees with that given in Reference 1.

In this paper we outline briefly our calculation[2] and compare and contrast it with the calculation in Reference 1. We later conclude that although both approaches start from the same initial Hamiltonian the treatments are drastically different and therefore it is not surprising that the conclusions differ. Of course the question which remains is, ''which calculation gives the more reliable estimate of the electron mass shift?'' We shall have something to say about that in our conclusions.

The starting point for nonrelativistic quantum electrodynamics is provided by the Hamiltonian

$$H = (1/2m_0)(\mathbf{p} - e\mathbf{A}_0)^2 + e\phi_0 + H_{\text{rad}} \equiv H_0 + H_I \tag{1}$$

where

$$H_0 = (\mathbf{p}^2/2m_0) + H_{\text{rad}} \tag{2}$$

and

$$H_I = -(e/m_0)\mathbf{p}\cdot\mathbf{A}_0 + (e^2/2m_0)\mathbf{A}^2 + e\phi_0 \tag{3}$$

H_0 gives the Hamiltonian for an electron in the radiation field in the absence of any interaction. H_I couples the finite-size electron, described by a fixed charge-density function $\rho(\mathbf{r})$, to the radiation field. The fields $\mathbf{A}_0$ and ϕ_0 are related to the electromagnetic fields $\mathbf{A}$ and ϕ by

$$\begin{aligned}
\mathbf{A}_0(\mathbf{r}) &= \int \rho(\mathbf{r} - \boldsymbol{\xi})\mathbf{A}(\boldsymbol{\xi})d^3r \\
\phi_0(\mathbf{r}) &= \int \rho(\mathbf{r} - \boldsymbol{\xi})\phi(\boldsymbol{\xi})d^3r
\end{aligned} \tag{4}$$

and they represent smeared or averaged fields at the center of the extended electron.

Using equation (1) together with standard commutation relations we can arrive at the Heisenberg equation of motion for the electron position operator. This equation is

$$m_0\ddot{\mathbf{r}} = e\mathbf{E}_0 + \tfrac{1}{2}(\dot{\mathbf{r}} \times \mathbf{B}_0 - \mathbf{B}_0 \times \dot{\mathbf{r}}) \tag{5}$$

In what follows we assume the presence of an external homogeneous field $\mathbf{E}_{0\text{ext}}$ as well the quantized electromagnetic fields $\mathbf{E}_{0\text{rad}}$ and $\mathbf{B}_{0\text{rad}}$. The equation of motion above is obviously independent of the gauge used. In our work we have chosen to use the Coulomb gauge while Moniz and Sharp have used the covariant or Lorentz gauge. The choice of a gauge is relevant to a certain degree since the fields **A**, ϕ also satisfy equations of motion, and these equations depend on the gauge. However, any final results should be invariant to the choice of gauge.

In our gauge, in the absence of any external magnetic field, **A** is transverse and satisfies an equation which relates it to the transverse current. We do not, however, solve this equation directly. Instead we make a Fourier expansion of this field at $t = 0$. The coefficients in the expansion are creation and annihilation operators of bare photons at $t = 0$. Our Hamiltonian and our state vectors will all be expressed at $t = 0$. In the Coulomb gauge a part of ϕ is due to the external electric field and another part is generated by the electron itself. We have shown earlier[3] that the part generated by the electron leads to an electrostatic interaction which shifts all energy levels in a state-independent way. It does not, therefore, contribute to the inertial mass shift which we discuss below.

Our procedure consists of taking the expectation value of equation (5) in a "physical" one-electron state. This is done at time $t = 0$. The expected value of the left-hand side will be proportional to the electron acceleration at $t = 0$. The expected value of the right-hand side will give the obvious term $e\mathbf{E}_{0\text{ext}}$, but in addition, as discussed below, the other terms will also yield contributions proportional to $\mathbf{E}_{0\text{ext}}$. We then define the inertial mass shift in the standard way as

$$m_{\text{phys}} = \frac{\mathbf{F}_{\text{ext}}}{\mathbf{a}} = \frac{e\mathbf{E}_{0\text{ext}}}{\langle \ddot{\mathbf{r}} \rangle} \tag{6}$$

Since the electron is assumed to be nonrelativistic, the expectation values of the force and acceleration should be related by Newton's second law. Our calculation of the inertial mass shift proceeds then from equation (6). Unfortunately, we do not know how to construct the exact physical one-electron state at $t = 0$, and therefore our procedure is to resort to perturbation theory. We construct a physical one-electron eigenstate

$$|\psi_{\text{in}} \rangle = \sum_n c_n \left(1 + \frac{1}{E_n - H_0 + i\epsilon} H_I \right.$$

$$\left. + \frac{1}{E_n - H_0 + i\epsilon} H_I \frac{1}{E_n - H_0 + i\epsilon} H_I + \cdots \right) |\psi_n\rangle \tag{7}$$

where $|\psi_n\rangle$ is an energy eigenstate of H_0 and $|\psi_{\text{in}}\rangle$ represents asymptotically a free slowly moving electron. The coefficients c_n are such that our state is a wave packet. The values of these coefficients have no bearing on the final result.

The acceleration can now be evaluated from

$$\mathbf{a} = \langle\psi_{in}|\ddot{\mathbf{r}}|\psi_{in}\rangle = (e/m_0)\langle\psi_{in}|\mathbf{E}_{0ext}|\psi_{in}\rangle + (e/m_0)\langle\psi_{in}|\mathbf{E}_{0rad}(\mathbf{r})|\psi_{in}\rangle$$
$$+ (e/2m_0)\langle\psi_{in}|[\dot{\mathbf{r}} \times \mathbf{B}_{0rad}(\mathbf{r}) - \mathbf{B}_{0rad}(\mathbf{r}) \times \dot{\mathbf{r}}]|\psi_{in}\rangle \quad (8)$$

We have evaluated the above keeping terms of order e^3 on the right-hand side of equation (8) and have found to this order that

$$\langle\psi_{in}|\mathbf{E}_{0rad}(\mathbf{r})|\psi_{in}\rangle \cong \frac{e^2}{3\pi^2 m_0} \int \frac{dk\,\bar{\rho}^2(\mathbf{k})}{(1 + k/2m_0)^2} \langle\psi_0|\nabla\phi(\mathbf{r})|\psi_0\rangle \quad (9)$$

and

$$\left\langle\psi_{in}\left|\frac{\dot{\mathbf{r}} \times \mathbf{B}_{0rad}(\mathbf{r}) - \mathbf{B}_{0rad}(\mathbf{r}) \times \dot{\mathbf{r}}}{2}\right|\psi_{in}\right\rangle$$
$$= \frac{e^2}{3\pi^2 m_0} \int \frac{dk\,\bar{\rho}^2(\mathbf{k})}{(1 + k/2m_0)^2} \frac{k}{2m_0} \langle\psi_0|\nabla\phi(\mathbf{r})|\psi_0\rangle \quad (10)$$

where $|\psi_0\rangle = \sum c_n|\psi_n\rangle$ and $\bar{\rho}(\mathbf{k})$ is the Fourier transform of $\rho(\mathbf{r})$.

To the accuracy we require, $|\psi_0\rangle$ may be replaced by $|\psi_{in}\rangle$ at this stage. We also rewrite $\nabla\phi(\mathbf{r}) = -\mathbf{E}_{ext}$.

From equations (8), (9), and (10) we find that the acceleration can be written as

$$\mathbf{a} = \frac{e}{m_0}\langle\psi_{in}|\mathbf{E}_{0ext}|\psi_{in}\rangle\left[1 - \frac{4\pi\alpha}{3\pi^2 m_0 c}\left(\int \frac{\hbar dk\,\bar{\rho}^2(k)}{[1 + (\hbar k/2m_0 c)]^2}\right.\right.$$
$$\left.\left.+ \int \frac{\hbar dk\,\bar{\rho}^2(k)(\hbar k/2m_0 c)}{[1 + (\hbar k/2m_0 c)]^2}\right)\right] \quad (11)$$

In this expression we have restored appropriate powers of $\hbar$ and c in order to be able to proceed to the classical limit. Using equation (6) we can now obtain the ratio $\delta m/m_0$ to leading order in α. We find

$$\frac{\delta m}{m_0} = \frac{4\alpha}{3\pi}\lambda_c\left[\int \frac{dk\,\bar{\rho}^2(k)}{(1 + \frac{1}{2}\lambda_c k)^2} + \int \frac{dk\,\bar{\rho}^2(k)}{(1 + \frac{1}{2}\lambda_c k)^2}\frac{1}{2}\lambda_c k\right] \quad (12)$$

where $\lambda_c = \hbar/m_0 c$ is the Compton wavelength of the electron. In equation (12) the first term is due to the electric force arising from $\mathbf{E}_{0rad}$ while the second term is due to the magnetic force implied by $\mathbf{B}_{0rad}$.

The result contained in equation (12) leads to the following conclusions:

1. For an extended electron the quantum-mechanical and classical results are finite.

2. For a point electron ($\bar{\rho} = 1$), quantum-mechanically the result diverges logarithmically.

3. For a point electron ($\bar{\rho} = 1$), classically (λ_c set to zero inside square brackets) the mass shift diverges linearly. It is worth noting that the second term in equation (12), which arises from the magnetic force, does not contribute classically.

4. The classical mass shift can be written as $\frac{4}{3}$ the electrostatic energy shift divided by c^2 where

$$E_{\text{electrostatic}} = \frac{\alpha \hbar c}{\pi} \int dk \, \bar{\rho}^2(k) = \frac{\alpha \hbar c}{2} \int d^3x \, d^3y \, \frac{\rho(\mathbf{x}) \, \rho(\mathbf{y})}{|\mathbf{x} - \mathbf{y}|} \qquad (13)$$

Note, however, that our *entire* mass shift arises from the interaction of electrons with the transverse part of the electromagnetic field, i.e., from its coupling to photons.

We will now try to explain how the Moniz and Sharp calculation proceeds. Their work is also based on equations (1) and (5), but instead of expanding the electromagnetic fields in a Fourier series at $t = 0$, they solve the inhomogeneous equation relating A_μ to j_μ, where A_μ is the electromagnetic field in the Lorentz gauge and j_μ is the electron current. An external field is not utilized.

The solution for A_μ contains an "in" field A_μ which is a solution of the homogeneous equation and which is an operator which can create and destroy "physical photons." It also contains a self-field $A_{\mu \text{ self}}$ which arises from the electron current, and which is related to it by the equation

$$A_{\mu \text{ self}}(\mathbf{r}, t) = \frac{e}{c} \int d^3 r' \, \frac{j_\mu(\mathbf{r}', t'_{\text{ret}})}{|\mathbf{r} - \mathbf{r}'|} \qquad (14)$$

If the right-hand side of equation (14) could be computed exactly it could be added to the "in" field and then substituted into equations (4) and (5). Instead of an exact calculation, which is of course highly unlikely, the authors of Reference 1 expand j_μ of equation (14) in a power series in $t'_{\text{ret}} - t$.

Two additional approximations have been made. These are:

1. Retain only those terms which will lead to a linear force on the right-hand side of equation (5). Thus any derivatives of $\mathbf{r}$ would be admissible but a contribution such as $\dot{r}^2 \mathbf{r}$ would not be retained. This is partially justified for an electron which is "almost" at rest.

2. The field $A_{\mu \text{ in}}$ has essentially been discarded. The expectation value of $A_{\mu \text{ in}}$ between "physical" one-electron states is zero. However, it does not follow that higher powers of this field and its derivatives will also have vanishing expectation values in this state.

After these approximations are made an equation of motion is obtained. This equation does not have a magnetic force term since such a contribution would be nonlinear. This conclusion is in contrast to ours; we find that both terms of the Lorentz force contribute to the mass shift even for a slowly moving electron. Moniz and Sharp then identify a mass shift δm which turns out to be entirely due to the electric field E_{self}. The mass shift obtained is finite and of order α when the charge density is extended. Two limits of this result can now be taken. If $\bar{\rho}$ approaches unity, δm vanishes. The classical limit of this result is zero. On the other hand, one can first proceed to the classical limit and then to the point limit. For the latter case the result is infinite and the divergence is linear.

Of course the interesting conclusion of the Moniz and Sharp work is the vanishing of the point-electron mass shift. This differs from our result, as given by equation (12). It is perhaps not surprising that the two calculations yield different mass shifts since the approximations used are quite different. Our result is perturbative and maintains quantization of the radiation field. It includes the entire electromagnetic field.

Since we have both started from the same model Hamiltonian there must surely be an answer to the question, "what is the electron mass shift?" Different approximation procedures will yield different results and therefore some will more reliably predict the "actual" δm than others. For an extended electron, our own preference is for the perturbation theory approach, but that's just an opinion which is to a large extent due to the many successes of perturbation theory in quantum electrodynamics.

References

1. E. J. Moniz and D. H. Sharp, *Phys. Rev. D* **10,** 1133 (1974); **15,** 2850 (1977).
2. H. Grotch and E. Kazes, *Phys. Rev. D* **16,** 3605 (1977).
3. H. Grotch and E. Kazes, *Phys. Rev. D* **13,** 2851 (1976).

12

Nonrelativistic Calculation of the Electron g Factor

H. Grotch and E. Kazes

Two of the major successes of modern quantum electrodynamics can be found in the calculations of the Lamb shift and the anomalous magnetic moment of the electron. Although no other theory can claim to successfully predict these quantities, it is nevertheless of interest to note that Bethe's approximately correct calculation of the Lamb shift was based on a nonrelativistic electron coupled to the electromagnetic field.[1] On the other hand, attempts to utilize the same model to compute $g - 2$ led to failure and didn't even result in the correct sign.[2]

In the present paper, which is a review of several previous publications,[3,4] we show that a positive value of $g - 2$ results when one takes into account mass renormalization. The result diverges, but for a reasonable cutoff the correct value is obtained. The extension to relativistic electron theory is then discussed.

The Hamiltonian for a nonrelativistic electron interacting with the electromagnetic field is given by

$$H = \frac{1}{2m_0} [\boldsymbol{\sigma} \cdot (\mathbf{p} - e\mathbf{A}_0 - e\mathbf{A}_{\mathrm{rad}})]^2 + H_{\mathrm{rad}}$$

$$= \frac{(\mathbf{p} - e\mathbf{A}_0)^2}{2m_0} - \frac{e}{2m_0} \boldsymbol{\sigma} \cdot \mathbf{B} - \frac{e}{m_0} (\mathbf{p} - e\mathbf{A}_0) \cdot \mathbf{A}_{\mathrm{rad}}$$

$$- \frac{e}{2m_0} \boldsymbol{\sigma} \cdot \mathbf{B}_{\mathrm{rad}} + H_{\mathrm{rad}} + \text{quadratic terms} \tag{1}$$

with $\mathbf{A}_0 = \frac{1}{2}\mathbf{B} \times \mathbf{r}$ and $\mathbf{A}_{\mathrm{rad}}$ being the quantized radiation field. H_{rad} is the free radiation field Hamiltonian while m_0 is the bare electron mass.

H. Grotch and E. Kazes • Department of Physics, Pennsylvania State University, University Park, Pennsylvania 16802

We can write this Hamiltonian as

$$H = H_0 - \frac{e}{2m_0}\boldsymbol{\sigma}\cdot\mathbf{B} + H_I \tag{2}$$

with H_I giving the interaction with the quantized radiation field $\mathbf{A}_{rad}$. This field may be expanded as

$$\mathbf{A}_{rad} = \frac{1}{(2\pi)^{3/2}}\int\frac{d^3k}{(2\omega)^{1/2}}\sum_\lambda\left[a(k,\lambda)\hat{\boldsymbol{\epsilon}}(k,\lambda)e^{i\mathbf{k}\cdot\mathbf{x}}\right.$$
$$\left.+ a^\dagger(k,\lambda)\hat{\boldsymbol{\epsilon}}(k,\lambda)e^{-i\mathbf{k}\cdot\mathbf{x}}\right] \tag{3}$$

There are several alternative ways of evaluating $g - 2$ based on the above Hamiltonian:

(1) We can evaluate $\langle\psi|H_{mag}|\psi\rangle$ where $|\psi\rangle$ is an eigenstate of H with $\mathbf{B} = 0$ and H_{mag} consisting of all terms in H which are linear in $\mathbf{B}$.

(2) We can treat the coupling to the radiation field as the perturbation, using as unperturbed states the electron "at rest" in the magnetic field.

(3) Similar to (2) except that the unperturbed state is chosen to be a Landau state. This method does not allow us to follow in detail the role of mass renormalization.

The first method was utilized in Reference 4 and therefore in the present work we will use the second method. The energy shift of the unperturbed state $|0\rangle$ is given by

$$\Delta E = \int\frac{d^3k}{2\omega(2\pi)^3}\langle 0|(-e/m_0)(\mathbf{p} - e\mathbf{A}_0)\cdot D^{-1}(-e/m_0)(\mathbf{p} - e\mathbf{A}_0)_\perp|0\rangle$$

$$+ \int\frac{d^3k}{2\omega(2\pi)^3}\langle 0|(-e/2m_0)(\boldsymbol{\sigma}\times\mathbf{k})\cdot D^{-1}(-e/2m_0)\boldsymbol{\sigma}\times\mathbf{k}|0\rangle$$

$$+\left\{\int\frac{d^3k}{2\omega(2\pi)^3}\langle 0|(-e/m_0)(\mathbf{p} - e\mathbf{A}_0)\cdot D^{-1}(-e/2m_0)\boldsymbol{\sigma}\times\mathbf{k}|0\rangle + cc\right\} \tag{4}$$

$$D = E_0 - \frac{e}{2m_0}\langle\boldsymbol{\sigma}\cdot\mathbf{B}\rangle - \frac{(\mathbf{p} - \mathbf{k} - e\mathbf{A}_0)^2}{2m_0} + \frac{e}{2m_0}\boldsymbol{\sigma}\cdot\mathbf{B}$$

Note the appearance of $\mathbf{k}$ in the square brackets. This arises because the exponential factors in equation (3) shift the momentum. Let us at first ignore this momentum shift. This corresponds to replacing factors such as e^{ikx} by 1 and is therefore the dipole or no-recoil approximation.

In this approximation only the second term of equation (4) gives a direct contribution to the magnetic moment. This term involves second-order perturbation theory with the coupling of the spin to the quantized magnetic field. The calculation is straightforward and leads to the energy shift

$$-\frac{e}{2m_0}\langle\boldsymbol{\sigma}\cdot\mathbf{B}\rangle\frac{\alpha}{2\pi}\left(-\frac{4}{3m_0^2}\int_0^\Lambda \omega\,d\omega\right) \tag{5}$$

Thus the *total* correction to the magnetic moment is *negative*. This can be understood by realizing that as the electron magnetic moment precesses about the external **B** field, the electron spin can be flipped momentarily owing to the virtual emission and reabsorption of photons. Thus on the average its projection along the external field direction is diminished, resulting in a smaller magnetic moment.

The reduction of the magnetic moment does not, however, imply a negative contribution to the g factor. The g factor is defined as the ratio of the magnetic moment to the "physical" Bohr magneton. Thus

$$g = \frac{|\boldsymbol{\mu}_0 + \delta\boldsymbol{\mu}|}{-(e/2m)} \tag{6}$$

where $\delta\boldsymbol{\mu}$ is the magnetic moment correction, which we have just found to be antiparallel to $\boldsymbol{\mu}_0$. The quantity m is the physical electron mass. This differs from the bare mass m_0 due to the interaction with the radiation field. Writing $m = m_0 + \delta m$ we have

$$\begin{aligned}
g &= \frac{|\boldsymbol{\mu}_0 + \delta\boldsymbol{\mu}|}{-(e/2m)} = \frac{|\boldsymbol{\mu}_0 + \delta\boldsymbol{\mu}|}{-(e/2m_0)}\left(1 + \frac{\delta m}{m_0}\right) \\
&\cong \frac{|\boldsymbol{\mu}_0|}{-(e/2m_0)} - \frac{|\delta\boldsymbol{\mu}|}{-(e/2m_0)} + \frac{|\boldsymbol{\mu}_0|}{-(e/2m_0)}\frac{\delta m}{m_0} \\
&= 2 - \frac{|\delta\boldsymbol{\mu}|}{-(e/2m_0)} + 2\frac{\delta m}{m_0}
\end{aligned} \tag{7}$$

Thus, although the magnetic moment decreases, mass renormalization gives a positive contribution to the g factor. The calculation of δm is straightforward. If we compute ΔE of equation (4) in a state in which $\mathbf{B} = 0$ and the electron has momentum $\mathbf{p}$, we find that the first term of equation (4) gives an energy shift

$$\frac{\mathbf{p}^2}{2m_0}\left(-\frac{4}{3}\frac{\alpha}{\pi}\int_0^\Lambda d\omega\right) \tag{8}$$

The total electron energy is therefore

$$\frac{\mathbf{p}^2}{2m_0}\left(1-\frac{4}{3}\frac{\alpha}{\pi}\int_0^\Lambda d\omega\right) \equiv \frac{\mathbf{p}^2}{2m} \tag{9}$$

which leads to the result $m = m_0 + \delta m$ with

$$\frac{\delta m}{m_0} \cong \frac{4\alpha}{3\pi}\frac{1}{m_0}\int_0^\Lambda d\omega \tag{10}$$

Combining equations (10), (7), and (5) we obtain

$$\frac{(g-2)/2}{\alpha/2\pi} \cong \frac{8}{3m}\int_0^\Lambda d\omega - \frac{4}{3m^2}\int_0^\Lambda \omega\,d\omega \tag{11}$$

To lowest order we may use either m or m_0 in equation (11).

This result diverges quadratically. For a cutoff $\Lambda = 0.42\ m$ the experimental g value is obtained. It is of interest to note that this is a sensible cutoff and that the calculation developed here would not have been reasonable if the cutoff required was so large that a relativistic theory was clearly needed.

Let us return to equation (4) and ask what happens if we do not make the dipole approximation. This calculation has been carried out[4] and it changes the cutoff required to obtain the correct result to $\Lambda = 0.53$ m. An interesting aspect of this calculation is that if the cutoff is removed, it leads to a perfectly finite but incorrect result of $(8/3)\,(\alpha/2\pi)$. However, it is clearly inconsistent to integrate to infinity when the electron is assumed to be nonrelativistic. This suggests carrying out the same calculation for the relativistic electron.

We have done this and, surprisingly, have found that the answer is identical to the nonrelativistic no-recoil calculation.[4] Apparently inclusion of relativistic effects cancels the recoil effects. We can see this cancellation occurring explicitly by adding onto the initial Hamiltonian higher-order terms in the Foldy–Wouthuysen reduction. Alternatively the calculation can be carried out in Dirac single-particle theory.[3]

For the latter case the energy shift replacing equation (4) will be

$$
\Delta E_n^{t\prime} = -\frac{e^2}{i} \int \frac{d^4k}{(2\pi)^4} \frac{1}{k^2 + i\epsilon}
$$

$$
\times \left\langle n \left| \alpha \cdot \left[\frac{\Lambda_+(\pi - k)}{E_n - k_0 - E(\pi - k) + i\epsilon} + \frac{\Lambda_-(\pi - k)}{E_n - k_0 + E(\pi - k) + i\epsilon} \right] \alpha_\perp \right| n \right\rangle
$$

$$
- \delta m_t' \langle \bar{n} | n \rangle \tag{12}
$$

where $|n\rangle$ represents the Dirac electron in a magnetic field and the operators Λ_+ and Λ_- are projection operators in the external field. In this calculation the mass shift is

$$
\frac{\delta m_t'}{m} = \frac{4}{3} \frac{\alpha}{\pi m} \int_0^\Lambda d\omega + \frac{\alpha}{\pi m^2} \int_0^\Lambda \omega \, d\omega \tag{13}
$$

The calculation has been carried out elsewhere[3] and leads to exactly the result given by equation (11). In the above expressions the symbol "t" stands for transverse since the entire contribution comes from the coupling of the electron to transverse photons. In the Coulomb gauge there is also an electrostatic energy resulting from the electrostatic self-interaction in the single-particle theory, but it turns out to merely shift all energy levels, independently of the particle momentum and of the external magnetic field by lim $a \to 0$ $\alpha/2a$. Hence there is no inertial mass shift and also no magnetic moment correction.

Finally we would like to mention that we have also carried out the full quantum electrodynamic calculation of $\alpha/2\pi$ in the Coulomb gauge. The relevant transverse and electrostatic energy shifts are given by

$$
\Delta E_n^t = -\frac{e^2}{i} \int \frac{d^4k}{(2\pi)^4} \frac{1}{k^2 + i\epsilon}
$$

$$
\times \left\langle n \left| \alpha \cdot \left[\frac{\Lambda_+(\pi - k)}{E_n - k_0 - E(\pi - k) + i\epsilon} + \frac{\Lambda_-(\pi - k)}{E_n - k_0 + E(\pi - k) - i\epsilon} \right] \alpha_\perp \right| n \right\rangle
$$

$$
- \delta m^t \langle \bar{n} | n \rangle \tag{14}
$$

$$
\Delta E_n^{es} = -\frac{e^2}{i} \int \frac{d^4k}{(2\pi)^4} \frac{1}{\mathbf{k}^2}
$$

$$
\times \left\langle n \left| \left[\frac{\Lambda_+(\pi - k)}{E_n - k_0 - E(\pi - k) + i\epsilon} + \frac{\Lambda_-(\pi - k)}{E_n - k_0 + E(\pi - k) - i\epsilon} \right] \right| n \right\rangle
$$

$$
- \delta m^{es} \langle \bar{n} | n \rangle \tag{15}
$$

TABLE 1. QED result for $\frac{1}{2}(g-2)/(\alpha/2\pi)$

Λ/m	Electrostatic	Transverse	Sum
2.04	3.221	−2.308	0.913
2.24	3.399	−2.477	0.922
3.16	4.099	−3.169	0.930
4.47	4.907	−3.929	0.984
10.00	6.884	−5.891	0.993
⋮			⋮
∞			1

Both of these expressions diverge although their sum is convergent. In carrying out the calculation we therefore regulate with $\Lambda^2/(\Lambda^2 - k^2 - i\epsilon)$ and later allow Λ to approach infinity. The above equations have factors of $-i\epsilon$ in the denominators which accompany the negative-energy projection operators, while the corresponding single-particle theory denominators would have $+i\epsilon$ instead. This makes an important difference for in the QED calculation the result is finite and, of course, correct. Table 1 illustrates the QED contributions as a function of Λ/m.

The nonrelativistic calculation carried out here was also done using Landau states. This calculation was first carried out by Arunasalam,[5] but his result contained a numerical error. We find that when this is corrected the result agrees with equation (11). It should be mentioned that in the special state chosen mass renormalization does not enter in an explicit manner since the energy of the unperturbed states is independent of the magnetic field.

We hope that this paper demonstrates that, based on the Hamiltonian given by equation (1), one can obtain the correct sign for the anomalous moment of the electron and even the right magnitude for a cutoff of 0.4 or 0.5 electron masses. Although these results seem to be qualitatively correct, as yet there is no substitute for the full QED calculation if one truly wishes to be quantitatively correct.

References

1. H. A. Bethe, *Phys. Rev.* **72**, 339 (1947).
2. T. A. Welton, *Phys. Rev.* **74**, 1157 (1948); Z. Koba, *Prog. Theor. Phys.* **4**, 319 (1949).
3. H. Grotch and E. Kazes, *Phys. Rev. D* **13**, 2851 (1976); see also the erratum [*Phys. Rev. D* **15**, 1184 (1977)].
4. H. Grotch and E. Kazes, *Am. J. Phys.* **45**, 618 (1977).
5. V. Arunasalam, *Am. J. Phys.* **37**, 877 (1969).

13

Fundamental Physical Problems of Quantum Electrodynamics

F. Rohrlich

1. Introduction

It is usually asserted that quantum electrodynamics (QED) is in excellent agreement with experiments and that the only troubles which beset this theory are of a mathematical nature: the removal of divergences by renormalization needs to be given a mathematically meaningful formulation.

While there is some truth in that, I do not quite share this opinion. I believe that our present formulation is in important respects *physically* incorrect and that an improvement in the physical description will bring with itself an improved mathematical structure. While, therefore, much has been written about the mathematical problems of QED, I wish to dwell briefly on those physical problems which go to the basis of the theory. These problems do not seem to be fully appreciated; I hope by this talk to encourage the study of some of them.

2. The Decomposition of the Electromagnetic Field

As an introductory remark it is worth recalling that already in the classical theory of point charges there are three *different* ways in which the electromagnetic fields can be separated into two parts.[1]

(A) The separation into irrotational (or longitudinal) and solenoidal

F. Rohrlich • Syracuse University, Syracuse, New York 13210.

(or transverse) fields is provided by the Coulomb gauge where φ yields the former, $\mathbf{A}$ the latter. It is not a covariant separation.

(B) The separation into velocity (generalized Coulomb) and acceleration (radiation) fields (or ρ^{-1} and ρ^{-2} fields where ρ is the invariant distance from the retarded source point) is relativistically invariant. Upon quantization only the acceleration fields are composed of physical photons. The "photons" of the velocity fields, if such are introduced, are unphysical and are eventually eliminated in the calculation. Unfortunately, Maxwell's equations cannot be separated covariantly into equations for velocity and for acceleration fields.

(C) The separation into bound and free fields, $F_+ = \frac{1}{2}(F_{\text{ret}} + F_{\text{adv}})$ and $F_- = \frac{1}{2}(F_{\text{ret}} - F_{\text{adv}})$, was used by Dirac in deriving the Lorentz–Dirac equation.[2] He called $2F_-$ "the radiation field"; it is asymptotically equal to the retarded radiation field of (B). This *does* permit covariant equations for each field separately. It also provides a neat separation into off-mass-shell and on-mass-shell fields when the theory is quantized.[3] Unfortunately, the physical meaning of F_+ and F_- is not intuitive because of the unconventional boundary conditions. We also know that it is not necessary to introduce these fields for the derivation of the Lorentz–Dirac equation. Purely retarded fields can serve this purpose too.[4]

The separations (A) and (B) are possible for any asymptotic condition, i.e., for any linear combination of retarded and advanced fields. In QED one can quantize the retarded transverse field (A) in the Coulomb gauge; it is equivalent to the quantized retarded radiation field (B) in the covariant gauge.

3. The Coulomb Problem

In conventional QED one does not do justice to the Coulomb interaction. This is so at least in part for technical reasons.

The perturbation solution of QED begins in zeroth order with a free electron characterized by a plane wave. But every physicist knows that this is an incorrect approximation whenever another charged particle (possibly with infinite mass, i.e., a Coulomb field) is present in the system: a Coulomb wave never becomes a plane wave asymptotically. A Coulomb interaction is very poorly approximated by an action on the electron at one instant of time with the particle being free before and after that interaction. Nor can a sum of repeated interactions of this type be expected to converge to the Coulomb interaction.

One should hasten to add that a treatment of QED in terms of

Coulomb propagators and Coulomb wave functions would complicate the computations drastically. And there lies the technical problem.

The price one pays for this incorrect approximation comes in various forms. One is in the infrared divergence problem. But there it enters "only" as a divergent phase factor which drops out of the cross section.[5] A well-known example should be recalled: Rutherford scattering in the Born approximation gives the *exact* result for the magnitude of the amplitude. The phase is incorrect but is not observable unless interference with nuclear scattering occurs. The contributions of all radiative corrections combine only to (divergent) renormalization factors of mass and charge, and to a correction of the phase.[6]

Another price one pays for the plane-wave approximation is the lack of a consistent embodiment of the bound-state problems into the *S*-matrix formulation. For example, the radiative corrections to bound-state energy levels cannot be computed without the extraneous addition of the unperturbed bound-state solutions to the results obtained by Feynman techniques.

But the most important deficiency lies in the way the Coulomb self-field is treated in conventional QED. Let us recall that according to the gauge invariance of the theory, QED in the Coulomb gauge gives in every respect the same results as QED in a covariant gauge. Then the contributions to the electron self-energy from the longitudinal field are neatly separated from the others as the contributions from the Coulomb potential in the Coulomb gauge.[7,8] The result is an infinite constant, linearly divergent,

$$m_1^{\text{coul}} \equiv \left\langle 1 \left| \frac{e_0^2}{8\pi} \int \frac{d^3x \, d^3y}{|\mathbf{x} - \mathbf{y}|} : \rho(\mathbf{x}, t): \, :\rho(\mathbf{y}, t): \right| 1 \right\rangle$$

essentially the same as for a classical point charge.

Now the above self-energy has exactly the structure that one obtains for a finite-size charge (rather than a point charge) when retardation is neglected. The usual argument is that retardation should not matter for a point charge. But in the point limit of a finite charge's self-energy with retardation, the retardation does contribute.

A power series expansion of retardation is an expansion in increasing orders of the time derivative, as is well known from purely classical calculations. But while one expands classically in powers of $R \partial/\partial t$, where R is a typical size parameter of the finite-size electron, the quantum-mechanical calculation[9] leads to an expansion in powers of $\lambda \partial/\partial t$, where λ is the Compton wavelength $\hbar/m$. In both cases one starts with a finite-size particle and takes the point limit at the end of the calculation.

But while the classical calculation involves only R, the quantum-mechanical calculation contains in addition the length λ which vanishes in the classical limit. And now one finds that the expansion in $\lambda \partial/\partial t$ gives a *completely different* result[9] from the expansion in $R \partial/\partial t$. Let the quantum-mechanical result be denoted by $m_2^{Coul}(R, \lambda)$ and the classical result, obtained in the $\lambda \to 0$ limit, by $m_2^{Coul}(R, 0)$. Neither results diverge. But while the point limit $R \to 0$ of $m_2^{Coul}(R, 0)$ diverges,

$$\lim_{R \to 0} m_2^{Coul}(R, 0) = m_1^{Coul} = \infty$$

the point limit of $m_2^{Coul}(R, \lambda)$ *vanishes,*

$$m_2^{Coul} \equiv \lim_{R \to 0} m_2^{Coul}(R, \lambda) = m_2^{Coul}(0, \lambda) = 0$$

This is the key result of Moniz and Sharp.[9,10] It agrees with my formulation of classical point electrodynamics[11] in giving a vanishing electromagnetic mass.

As a side remark one learns from this that a classical point charge should *not* be considered to be the point limit of an extended classical charge, rather it should be considered to be the classical limit of a point *quantum-mechanical* particle. The two limits $R \to 0$ and $\lambda \to 0$ are not interchangeable.

The difference between m_1^{Coul} and m_2^{Coul} is retardation. Since this is obtained from an expansion in *positive* powers of $\hbar$, viz., $\lambda(\partial/\partial t)$, it cannot possibly result from perturbation expansion which is in *negative* powers of $\hbar$, viz., in powers of α. Note[12] that the leading term of $m_2^{Coul}(R, \lambda)$ is identical with m_1^{Coul}.

Two objections can be raised against this result. One is that reference 9 is using the Lorentz gauge instead of the Coulomb gauge. But the theory is gauge invariant and in fact both references 8 and 9 start with the same gauge invariant equation of motion. The second objection refers to the neglect of terms nonlinear in 'the velocity. This is justified by the nonrelativistic nature of these calculations.

There are of course non-Coulombic contributions to the self-energy. They combine with the Coulombic ones to form an overall relativistic self-energy which is logarithmically divergent according to conventional QED. To a certain approximation these were computed by Grotch and Kazes.[8]

One concludes from this study that conventional QED does not treat the Coulomb interaction in satisfactory approximation and, in particular, one seems to ignore the associated retardation effects which are essential in view of the above example.

4. The Infrared Divergence Problem

This problem is as old as quantum electrodynamics. It was "solved" for the first time[13] in 1937 and has been "solved" since then many times by many people. With the very recent work by Zwanziger[14] we have come very close to a complete solution. In view of his presentation there I want to restrict myself to the following short comments.

The infrared divergent contributions from the velocity field and the acceleration field can be covariantly separated. They appear as "charge photons" and "radiation photons" in his work. One suspects that a nonperturbative treatment of the Coulomb field (in a covariant way) will make the "charge photons" unnecessary and will restrict the problem to transverse soft quanta (radiation photons). One lesson learned over the years is that these, too, cannot be treated in the conventional (Fock space) perturbation method. One needs coherent states or some other representation of the commutation relations.

We have also learned that the physical electron is a very complicated structure whose surrounding "photon cloud" has yet to be described exactly. Such a description is necessarily part of a complete solution of this problem. Another characteristic feature of a correct physical description is scattering cross sections of charged particles which *vanish* in the exactly elastic limit. This has already been accomplished to a certain extent.

I consider the problem of the correct description of a physical electron the key to a QED which is free of infrared divergences. Not knowing any better we are at present willing to be led there by mathematical considerations.[15]

Through these considerations one also comes to the conclusion that conventional QED is incorrect. The Hilbert space generated from the vacuum by the in-fields ψ^{in} and A^{in} cannot contain interacting states unless ψ^{in} and A^{in} do *not* commute, contrary to what is assumed in usual QED. What is needed for a correct description of an electron are non-Fock electromagnetic states.

5. The Classical Limit Problem

Most physicists probably subscribe to the notion that classical (non-quantum) electrodynamics (CED) can be derived from QED by a suitable limiting process. This notion is logically very reasonable and in my view philosophically necessary because QED without CED is incomplete. This is easily seen by analyzing a measurement process in QED. And yet, there does not exist to date a clean proof of this limit. The problem lies of course to a large extent in defining the classical limit appropriately.

In pure QED electrons are structureless and must therefore have classical point charges as limits. The CED of these particles is defined by the simultaneous set of Maxwell–Lorentz field equations and the Lorentz–Dirac equation, presumably with a suitable asymptotic condition.[11] It is not very difficult to obtain the field equations with external sources as a limit; the problem lies with deriving the *simultaneous* set of equations where both the particle and the field are treated dynamically. In particular there is the problem of deriving the Lorentz–Dirac equation,

$$m\dot{v}^{\mu} = F^{\mu}_{\text{ext}} + \frac{e^2}{6\pi}\,(\ddot{v}^{\mu} - \dot{v}^2 v^{\mu})$$

In view of the widespread belief that there are serious objections to this equation, I want to intersperse here a few remarks about these objections, the runaway solutions, and the preacceleration (noncausal motion).

One must keep in mind what I emphasized already once before: the classical description is meaningful only to the extent that it is the limit of a quantum-mechanical description. This limit determines whether these difficulties are or are not present. At this point the work of Moniz and Sharp is relevant once more. They prove that in the point limit of the extended quantum-mechanical charge the equations of motion permit neither runaway solutions nor preacceleration.[9] Theirs is a quantum mechanical argument. A *classical* argument to that effect would necessarily be at least partly *ad hoc*. (This result gives additional credence to the work of Moniz and Sharp and provides further support for the consistency of their approximation of neglecting certain nonlinear terms.)

The physical origin of the various terms in the Lorentz–Dirac equation should be clear. The term with parenthesis is the Laue four-vector of "radiation reaction." It consists of two terms, the Schott term containing a $\ddot{v}$ and the fourvector of the rate of emission of radiation four-momentum. The latter is the covariant generalization of the Larmor formula and is the true radiation reaction term; it originates entirely in the acceleration fields. The somewhat mysterious Schott term results from the interference between velocity fields and acceleration fields. This comes from the fact that the energy tensor is bilinear in the fields. The details can be seen from the work of Teitelboim.[4]

In the relativistic case, the classical limit singles out of the electron field one electron of positive energy. The problems one faces have to do with the following questions: What becomes of the pair production effects (for example, virtual creation and annihilation) in the classical limit? What becomes of radiative corrections in the classical limit? What becomes of the soft photon emission that necessarily accompanies all interactions of a charged particle not in a bound state?

This latter problem is especially interesting because the infrared problem has a certain *similarity* to the classical limit problem: in both instances the number of photons is very large such that the photon number concept is no longer meaningful. It is therefore not surprising that all indications lead to a dominant role for the coherent-state representation also for the classical limit problem.[16-18] The correspondence principle can be adapted especially to this problem if it is cast into the coherent-state language.[19]

There are of course also essential *differences* between the physical situations of the infrared problem and the classical limit. In the former the current is assumed unaffected by the radiation emission, in the latter case the momentum and energy loss is essential and contributes exactly the last term in the Lorentz–Dirac equation. The Schott term is always concomitant with radiation emission because without the $\ddot{v}$ term the Lorentz–Dirac equation would be mathematically inconsistent: since $\dot{v} \cdot v = 0 = F_{ext} v$, it would imply $\dot{v}^2 = 0$, i.e., $v^\mu = $ const., a free particle.

6. The Closure Problem

For many years now it has been a trite observation that there are four kinds of fundamental interactions of vastly different strengths of which the electromagnetic one is the best understood. It is therefore reasonable to attempt to formulate QED as a self-contained theory independent of the other interactions. One is thereby willing to accept the masses and charges of the fundamental particles (e.g., the electron or the muon) as phenomenologically given. If such a formulation were possible, QED would be called a closed theory.

However, there has long been evidence in the literature that QED is not closed, i.e., that a renormalizable QED is not possible without further assumptions if one tries to formulate this theory for π-mesons or other spinless particles. The need for further assumptions can be thought of as indicating a cutoff between QED and other (nuclear) interactions.

In 1950 I found that the QED of spin-zero particles is not renormalizable because a particular physical process (meson–meson scattering) remains divergent even after mass and charge renormalization.[20] The remedy that suggests itself immediately is a new, nonelectromagnetic interaction of the form $\lambda\phi^4$ with an unrenormalized coupling constant λ. One can then remove the objectionable scattering terms by a renormalization of λ. The result is an additional (necessarily observable) interaction with renormalized coupling constant λ_{ren}. It would be meaningless and counter to the spirit of renormalization theory to put $\lambda_{\text{ren}} = 0$ by fiat.

The measurement of λ_{ren} is made difficult by the presence of nuclear π–π interactions and by the question of the extent to which the $\lambda\phi^4$ term is nothing but a nuclear interaction term. This situation is therefore evidence for the claim that QED is not closed. It must be considered together with nuclear interactions in order to provide a finite description of π-mesons (or any spinless charged particles).

The development of the Weinberg–Salam theory in recent years has provided a unification of electromagnetic and weak interactions. But this unification has so far not provided a predicted value for λ_{ren}, nor has this problem been resolved in some other way.

7. Closing Remarks

We have not exhausted the list of fundamental physical problems. For example we did not discuss two problems which are perhaps even more difficult and whose solution may lie even farther in the future. One is the mass spectrum problem of the leptons of which only four levels are known so far, ν, e, μ, and the recently discovered τ. The other is the problem of the fundamental charge. Given the minimum empirical data, i.e., any three data that determine the scales of length, time, and mass such as m_e, $\hbar$, and c, all other physical quantities should be derivable from theory. With units in which these are chosen as 1, why is the fundamental charge $e = (137)^{-1/2}$?

Our very brief discussion of these fundamental problems is presented here in the hope that it will stimulate research on some of them. Only the deeper insight which one can expect to gain by such work will enable us to improve the present formulation of QED on physical grounds. The concomitant improvement in the mathematical structure can then be expected either as an act of faith or as an educated prediction.

References and Notes

1. A. O. Barut, *Electrodynamics and Classical Theory of Fields and Particles*, MacMillan, New York (1964); F. Rohrlich, *Classical Charged Particles*, Addision-Wesley, Reading, Massachusetts (1965).
2. P. A. M. Dirac, *Proc. R. Soc. London Ser. A* **167**, 148 (1938).

3. F. Rohrlich, *Acta Phys. Austriaca Suppl.* **XIII**, 487 (1974), and The nonlocal nature of electromagnetic interactions in *Physical Reality and Mathematical Description* Eds. C. Enz and J. Mehra, Reidel, Holland (1974).

4. C. Teitelboim, *Phys. Rev. D* **1**, 1572 (1970) and **3**, 297 (1971).

5. J. M. Jauch and F. Rohrlich, *Theory of Photons and Electrons*, 2nd ed., Springer-Verlag, New York (1976), Supplement 4.

6. R. H. Dalitz, *Proc. R. Soc. London Ser. A* **206**, 509 (1951), reference 5, p. 330.

7. J. D. Bjorken and S. D. Drell, *Relativistic Quantum Fields*, McGraw-Hill, New York (1965), pp. 88 and 92.

8. H. Grotch and E. Kazes, *Phys. Rev. D* **13**, 2851 (1976) and **16**, 3605 (1977). The latter paper is discussed by these authors in Chapter 11 of this volume.

9. E. J. Moniz and D. H. Sharp, *Phys. Rev. D* **15**, 2850 (1977) and **10**, 1133 (1974). This paper is discussed by the second author in Chapter 16 of this volume.

10. F. Rohrlich, *Acta Phys. Austriaca* **41**, 375 (1975).

11. F. Rohrlich, *Phys. Rev. Lett.* **12**, 375 (1964), and reference 1, F. Rohrlich, Sections 6–9 and 7–1. See also J. D. Hamilton, *Am. J. Phys.* **39**, 1172 (1971).

12. Compare equation (3.19) of the first paper cited in Reference 9. A factor 4/3 there would be absent in a relativistically consistent calculation.

13. F. Bloch and A. Nordsieck, *Phys. Rev.* **52**, 54 (1937). For later papers see reference 5, Section 16-1 and Supplement 4.

14. D. Zwanziger, Infrared catastrophe averted by Hertz potential, *Phys. Rev. D* **19**, 473 (1979).

15. J. Fröhlich, G. Morchio, and F. Strocchi, Charged sectors and scattering states in QED, *Ann. Phys.* **119**, 241 (1979).

16. J. R. Klauder, *J. Math. Phys.* **4**, 1058 (1963) and **5**, 177 (1964).

17. I. Bialynicki-Birula, *Ann. Phys.* **67**, 252 (1971); *Acta Phys. Austriaca Suppl.* **XVIII**, 111 (1977).

18. K. Hepp, *Commun. Math. Phys.* **35**, 265 (1974).

19. J. R. Klauder, *J. Math. Phys.* **8**, 2392 (1967).

20. F. Rohrlich, *Phys. Rev.* **80**, 666 (1950).

14

Electromagnetic Interactions beyond Quantum Electrodynamics

A. O. Barut

1. Introduction

There is growing evidence that the electromagnetic interactions of fundamental particles have a much richer and deeper structure than that given by the perturbation series of quantum electrodynamics. It is the purpose of this paper to discuss the directions in which either the methods or the concepts of quantum electrodynamics have to be generalized and to point out possible novel and unexpected phenomena.

2. Generalizations in Methods

First, adhering precisely to the formalism of QED, are there areas where the scope of the theory can be generalized?

2.1. Ambiguities in Renormalization: The μ-Meson

It is generally believed, with few exceptions,[1,2] that the renormalization procedures of QED are "unambiguous," even if the theory is not quite satisfactory mathematically because of divergences. I shall therefore point out an ambiguity in the idea of renormalization which I believe also has important physical consequences.

In order to carry out the renormalization we have to know to what

A. O. Barut • Department of Physics, The University of Colorado, Boulder, Colorado 80309

final physical situation we renormalize our theory. We must ask: "Renormalization to what?" We tacitly assume that we renormalize it to the final physical observed electron and photon. We demand, for example, that the fermionic propagator of the interacting theory look like a free fermionic propagator, within a factor, at the renormalization point. This demand then defines renormalized masses and renormalized fermion and wave functions by (infinite) factors times the unrenormalized quantities. Similarly, the photon propagator and the vertex functions are renormalized to their corresponding free forms. We can, however, renormalize the theory to a different final situation, for example, to two fermions, the electron with renormalized mass M_e, and the muon, with renormalized mass M_μ, in which case the muon can be considered as a kind of excited state of the electron (or, e and μ as two distinct excitations of the underlying bare fields). The fermionic propagator in this case has to be written in such a way that it has two poles corresponding to masses M_e and M_μ. The renormalization prescriptions will be different than the usual prescriptions, but the successful results of renormalized QED will not change. It has been recently shown[3] that a generalized unified QED based on the two- (or more-) mass fermion equation,

$$(-i\gamma^\mu D_\mu - m_1)(-i\gamma^\mu D_\mu - m_2)\Psi = 0 \qquad D_\mu = \partial_\mu - ieA_\mu \qquad (1)$$

is renormalizable. This theory accounts for equal charge and identical electromagnetic interactions of two fermions separately and leads naturally to a superselection rule between the different fermionic subspaces so that the decay $\mu \to e + \gamma$ does not occur. The indefinite metric does not cause any problem in such a theory with a superselection rule. In each fermion sector we can use the standard perturbation theory. However, the pair $(\mu^+\mu^-)$ again belongs to the Fock space of the electron sector with positive metric. Consequently, the usual QED is enlarged by the μ-pair production processes.

The physical picture of how the μ-meson arises from electromagnetism may be stated as follows. The radiative effects induce an anomalous magnetic moment in the electron. This in turn couples with the self-field of the electron giving an extra mass which is then quantized. Indeed, the radiative effects in classical electrodynamics give a closed expression (see Sections 2.2 and 3) for the classical anomalous magnetic amount of the electron, $g = 2(2\alpha/3)$.[4] If we write a mass quantization equation in the form

$$(-i\gamma^\mu \partial_\mu + \lambda \partial_\mu \partial^\mu + \kappa)\Psi = 0 \qquad (2)$$

generalizing the Dirac equation, and determine the two constants λ and κ by the values of M_e and $g = 4\alpha/3$, then the equation predicts a second mass M_μ with

$$M_\mu = M_e(3/2\alpha + 1) \tag{3}$$

This formula clearly indicates an electromagnetic nature of the μ-meson, as it depends on α alone.

Equation (3) can be generalized to the τ-meson and other possible higher excitations of the lepton. Semiclassically the magnetic excitation energy of a system consisting of a charged particle and a magnetic moment is quantized in mass quanta of the form λn^4, where n is a new quantum number and $\lambda = 3/2\alpha$,[5] If we postulate that in the sequence e, μ, τ, . . . this quantized mass is added to the rest mass we obtain for $n = 1$ the equation (3), and for $n = 2$, the mass

$$M_\tau = M_e + M_e(3/2\alpha) + M_e(3/2\alpha)2^4 = 1786.08 \text{ MeV} \tag{4}$$

The most recent experimental value is $M_\tau = 1787 \pm 10$ MeV. The mass predicted for the next excitation ($n = 3$) would be $M_\delta = 10.293$ GeV. It is to be expected that a complete nonperturbative treatment of radiative effects would lead to these phenomena.

Renormalization for Bound-State Problems

That the propagator of a particle can contain other poles has in fact other precedents. For example, the photon propagator in QED must contain infinitely many poles corresponding to positronium states (even many more, see Section 2.2). Again, these poles cannot be obtained in any finite order of perturbation theory. But they can be seen, for example, in the eikonal approximation.[6] An infinite-component wave equation, generalizing (1), contains these infinitely many poles coupled to the photon.[7] In the case of bound-state problems in QED, a Schrödinger or Dirac-type wave equation with Coulomb potential is usually taken as a starting point for the calculation of radiative corrections to the bound states. In this case the potential picture, in fact, can be viewed as a new renormalization procedure, knowing that we have to renormalize the theory to a final solution with infinitely many mass states.

2.2. Nonperturbative Phenomena

The success of quantum electrodynamics in simple scattering problems has given the impression that electromagnetic effects between charged particles are adequately taken into account by one- (or few-) photon exchange diagrams because of the smallness of the coupling constant. In fact this is the standard procedure in high-energy particle physics: Effects not described by essentially one-photon exchange diagrams are attributed to or lumped into other (strong or weak) interactions. While there are certain states of the system in which the expectation value of the interaction is small, hence the perturbation theory is reasonable, there are also other *new* states or solutions of the total Hamiltonian in which the expectation value of the interaction is very large. These states are "far away" from the perturbative solutions and cannot be seen at all in any order of perturbation theory. Since this point is central to the discussion of this section, we present here a simple model version of the nonperturbative phenomenon that we are talking about.

Consider the magnetic interactions of two charged particles in ordinary quantum mechanics. The reduced radial equation in the center-of-mass frame can be brought to the form

$$\left(\frac{d^2}{dr^2} + k^2 + \frac{\alpha}{r} + \frac{A}{r^2} + \frac{B}{r^3} + \frac{C}{r^4}\right) u = 0 \tag{5}$$

The last three terms come from the magnetic vector potential and angular momentum barrier while α/r is the Coulomb part. The constants A, B, C are, in general, dependent on energy and angular momentum. If we first solve the Coulomb part with stationary energies E_0 and solutions u_0 and treat the last three terms as a perturbation V_1, then indeed in such states the quantity $\langle u_0| V_1|u_0\rangle$ is small, corresponding to the well-known fact that magnetic effects are small in atomic processes. However, if we plot the potential V_1, we obtain a deep potential well and realize that equation (5) can have entirely new resonance states u_1 of large positive energies in which the wave function is localized at much shorter distances than the Coulomb wave functions, and we find indeed that the quantities $\langle u_1| V_1|u_1\rangle$ in these new states are now very large. In fact, the states u_1 are entirely caused by V_1 and for such states the Coulomb potential $V_0 = \alpha/r$ may be treated as a perturbation. This is a truly nonperturbative effect. The perturbation theory around E_0, using the complete set of states of u_0, is completely inadequate for the description of these new states. This phenomenon may be demonstrated in many models of magnetic interactions.[8]

2.2.1. Iteration Instead of Perturbation

A method well-suited to describe the nonperturbative states described above from first principle is the following. We start from the coupled Maxwell–Dirac equations. Eliminating the potentials A_μ, we may write a nonlinear integrodifferential equation for the Dirac field,

$$(-i\gamma^\mu \partial_\mu - m)\Psi(\mathbf{x})$$

$$= e^2 \gamma^\mu \Psi(\mathbf{x}) \int d\mathbf{y}\, D(\mathbf{x} - \mathbf{y})\bar{\Psi}(\mathbf{y})\gamma_\mu \Psi(\mathbf{y}) + e\gamma^\mu \Psi(\mathbf{x})A_\mu^{\text{in}}(\mathbf{x}) \quad (6)$$

where A_μ^{in} or A_μ^{ext} comes from the solution of Maxwell's equations, and $D(\mathbf{x} - \mathbf{y})$ is the Green's function for the electromagnetic field with suitable boundary conditions. Equation (6) can be used both in first- and second-quantized versions. In the first case, we may write similar coupled equations for a system of two (or more) particles with wave functions $\psi_1(\mathbf{x}_1)$, $\psi_2(\mathbf{x}_2)$, ... The idea is now to find directly bound states and resonance solutions of (6) by iteration: We insert a trial localized solution on the right-hand side, evaluate the integral, and solve the Dirac equation in the resultant potential. It is possible in this way to obtain terms corresponding to the magnetic interactions described above.[9,10]

The approach parallels the treatment of radiative effects in classical electrodynamics. It is remarkable that radiation reaction can be obtained in classical electrodynamics nonperturbatively and in closed form. Also the renormalization process of (6) can be carried out parallel to the classical theory. Even if the theory is finite [which is the case if one works always with localized states for which the integral in (6) exists] a renormalization is necessary because the self-energy in (6) is already counted in the mass term: When $A_\mu^{\text{in}} \to 0$ (i.e., free particle) we must regain the free Dirac equation with experimental mass m.

Another advantage of (6) is that we have now very direct limits to various approximations of the two-body problem, such as the Breit–Pauli and Dirac equations.

2.2.2. The Richness of Electrodynamics

With these nonperturbative resonance states quantum electrodynamics suddenly becomes enormously richer than it was thought to be, even beyond the phenomena of the μ- and τ-mesons discussed in Section 2.1.

Starting with the three absolutely stable particles, the proton, the electron, and the neutrino and their antiparticles, it is possible to identify all hadron states with the magnetic resonance states of electrodynamics. In fact all multiplets of hadrons described by the internal groups $SU(3)$ [or $SU(4)$, . . .] have representations in terms of the magnetic states of p, e, and ν.[11] The assumptions are that μ (τ, . . .) are excitations of the electron, and that $\nu, \bar{\nu}$ possess an anomalous magnetic moment (as a limit of the Dirac particle with $e \to 0$, $m \to 0$, $e/m \to \kappa$).

Furthermore, the hadron resonances decay via barrier penetration and the rearrangement of constituents or via the μ-decay. Consequently weak interactions find also a place in a nonperturbative electrodynamics; they are much like the α-decay.

One might argue that we don't seem to see strong interactions between p and e, or e^+e^-, or $e\nu$, that is why strong and weak interactions have been introduced in the first place. This comes about as follows. The essential ingredient of magnetic resonances is a strong potential barrier between the constituents followed by a potential well. In such a situation the cross section for resonance penetration is extremely small and must be measured in careful, new predicted experiments. One should look, for example, for small bumps in the cross sections of e^+e^- or $e^-\mu^+$ around the π^0 or K^0 masses. Once, however, the stable particles are paired into mesons and baryons, they can interact strongly if there is no potential barrier via magnetic interactions and via the formation and recombination of further lepton pairs.[12]

These results may indicate that electrodynamics in its widest scope is an *already-unified* theory, with no further arbitrary coupling constants, or no new fields and no new interactions.

3. Generalizations in Concept

Besides generalizations in the method of standard QED perturbation theory we might contemplate the generalization of the notion of the electron and the electron theory itself. The reason for this lies in the nature of the fine-structure constant α and in the idea of the electromagnetic origin of the mass m of the electron. "The theoretical determination of α is certainly the most important of the unsolved problems of modern theoretical physics".[13]

In quantum electrodynamics both α and m are primary inputs and therefore cannot be calculated. I believe that the understanding of these concepts and hence of the quantum theory itself necessitates a generalization of the description of the structure of the electron. As the simple-minded world-line picture of the electron is approximate, so is its description by a wave function or a one-particle state in QED. Any indication towards such a generalization should therefore be of considerable interest. A clue might come from the process of emission and absorption of photons from an electron, where quantum theory itself originated. QED treats these as primary acts of creation and annihilation not to be analyzed further. Creation and annihilation operators are introduced which just create or annihilate photons of energy momentum $\hbar k$. I think we can enlarge the scope of QED if we try to analyze the emission process of a photon as a kind of radiation from a charged particle. This amounts of course to analyzing the origin of the Planck–Einstein relation $E = \hbar\omega$ for a photon instead of accepting it as a first principle. In classical electrodynamics the energy radiated by an accelerating charge in one period of oscillation goes like $E = \lambda\omega^3$ and not like $E = \hbar\omega$, as in the case of quantal photon radiation. However, we can understand the linear dependence of E on ω if we give to the motion of the charge a structure like the *Zitterbewegung* known from the Dirac equation. According to the Dirac equation there is a difference between the position operator x of the charge and the position operator X of the center-of-mass of the electron. The latter is moving like a relativistic particle of mass m should, but the former performs an erratic motion with the velocity $\dot{x} = c\alpha$, hence with eigenvalues $\pm\,c$ for each component of the velocity. The acceleration of the charge is $\ddot{x} = c(d\alpha/dt)$. For the radiation emitted by the charge we should use of course $\ddot{x}$ and not $\ddot{X}$. I have shown else-where[14] that if we use expectation values of $\ddot{x}$, we obtain the Einstein A coefficient for spontaneous emission of radiation, whereas if we use the *eigenvalues* of $\ddot{x}$, we obtain a linear radiation formula $E \sim \omega$. The proportionality constant under certain assumptions gives the numerical value of the Planck constant $\hbar$. This does not mean a return to a naive classical particle model. However, it is possible that an unusual structure or motion of the charge could both lead to probability amplitudes or wave functions and to Planck's formula. If this path is taken then the electro-magnetic concept of mass as the energy in the near field of the charge assumes an entirely new dimension, and so do the self-energy singulari-ties of QED.

It would be very valuable to study in this connection the solutions of the classical nonlinear Lorentz–Dirac equations with radiation reaction and with spin because these equations include, as we have noted, the radiative phenomena nonperturbatively and in closed form. Already in classical theory one may obtain a structure for the electron which is neither a point nor an extended object, but something like the Zitterbew-egung of the Dirac electron, or even more.

References

1. P. A. M. Dirac, *Europhysics News* **8**, 1 (1977) and *Proceedings of the Conference on Particle Physics, Budapest 1977.*
2. F. Rohrlich, Chapter 13 of this volume.
3. A. O. Barut and J. P. Crawford, *Phys. Lett. B.* **82**, 233 (1979).
4. A. O. Barut, *Phys. Lett. B* **73**, 310 (1978).
5. A. O. Barut, Lepton mass formula, *Phys. Rev. Lett.* **42**, 1251 (1979).
6. A. O. Barut and Z. Z. Aydin, *Nucl. Phys. B* **41**, 150 (1972).

7. A. O. Barut and A. Baiquni, *Phys. Rev.* **184**, 1342 (1967).

8. A. O. Barut and J. Kraus, *Phys. Lett. B* **59**, 175 (1975); *J. Math Phys.* **17**, 506 (1976).

9. A. O. Barut and J. Kraus, *Phys. Rev. D* **16**, 161 (1977).

10. A. O. Barut and R. Raczka, *Acta Phys. Pol. B.* (August 1979).

11. A. O. Barut, *Proceedings of the Texas Conference on Group Theoretical Methods and Mathematical Physics, Lecture Notes in Physics,* Vol. 94, Springer-Verlag, New York (1979), pp. 490–498.

12. A. O. Barut, *Symposium on Symmetries in Science,* Ed. B. Gruber, Plenum Press, New York (1980), and *Surveys in High Energy Physics*, Vol. 1, No. 2 (1980).

13. W. Pauli, in *Albert Einstein: Philosopher–Scientist,* Ed. P. A. Schilp, Library of Living Philosophers, New York (1949), p. 158.

14. A. O. Barut, *Z. Naturforsch. Teil A* **33**, 993 (1978); *Czech. J. Phys. B* **29**, 3 (1979); in *Quantum Theory and the Structures of Time and Space*, Vol. III, Ed. L. Castell and C. F. von Weizsäcker, C. Hanser-Verlag, Munich (1979), pp. 175–182.

15

Macroscopic Quantum Electrodynamics

Kimball A. Milton

1. Introduction

Source theory[1] is an effective, nonspeculative attitude and framework in which to treat the interactions of quantum particles. It is a field theory without operators in which, as in classical field theory, phenomenological fields are defined in terms of the response of the action to probe sources. The transition to classical physics is thus smooth. It is an effective approach in that renormalization, which connects operator fields to their particle content, does not occur. It is effective also in the practical sense: a great variety of calculations in gravitational, weak, electromagnetic, and strong interactions can be carried out simply and with physical clarity.

One of the successes of the source theory attitude toward physics lies in the domain of macroscopic quantum electrodynamics. The questions treated hitherto fall into two categories:

(i) First are those having to do with the motion of charged particles in external homogeneous magnetic (electric) fields. Essentially, the solution of such problems employs, or effectively generates, a proper time representation for the electron Green's function in a field of arbitrary strength. The latter was first derived by Schwinger in his famous 1951 paper,[2] which evidently already embodies much of the modern source-theoretical point of view. Among the applications of this exact Green's function technique are recalculations of synchrotron radiation[3-8] includ-

Kimball A. Milton • Department of Physics, University of California, Los Angeles, California 90024. Supported in part by the National Science Foundation.

ing the quantum correction[9,10] and the first reliable calculations of electron[11-13] and neutrino[14] pair creation by photons moving in strong fields (including, in the former case, the resulting index of refraction and attenuation coefficient), Compton scattering,[15,16] electron pair creation by photon–photon scattering,[17] and the new phenomenon of synergic synchrotron-Cerenkov radiation.[18] The latter has evident laboratory applications, while the other processes have important astrophysical implications, especially in pulsar physics.

(ii) The second concentration of effort is less fully developed. It has to do with the Casimir effect,[19] which is the vacuum quantum-electrodynamic force between bulk dielectrics and conductors and is to be identified with the van der Waals force between molecules.[20] In contrast to the above, a source theory treatment is based on the photon Green's function subject to the macroscopic classical boundary condition on the surfaces.[21] The theoretical development here is less advanced in that no quantum corrections to the photon Green's function have yet been considered. Applications include rederivations of the Casimir forces between parallel-plane dielectric surfaces and conducting planes[22] and of the Casimir energy of a conducting spherical shell[23]; a new, suggestive result indicates that the latent heat and surface tension of a liquid like liquid He can be understood in large part as a manifestation of macroscopic quantum electrodynamics.[22]

These subjects are already too vast to survey in detail in a short review. Moreover, it must be borne in mind that in each particular problem special methods and techniques are developed—we have not proceeded by following rote procedures. It will be necessary therefore to be selective in this outline and to emphasize some of the most essential physical ideas that enter into the calculations. For specific details, the reader is referred to the original literature, for which a complete bibliography is supplied.

2. Motion of Charged Particles in Strong Electromagnetic Fields

It is possible to make great progress in describing quantum-electrodynamic processes in homogeneous electromagnetic fields because the equations of motion are then integrable, being essentially equivalent to the harmonic oscillator problem. As a consequence we can obtain results exact in the field strength, although we are still limited to a perturbative expansion in the fine-structure constant α.

The methods of attacking such problems fall into two broad categories. In the first, one makes use of the exact electron Green's function, first obtained in 1951.[2] This may be written as

$$G(x, x') = \Phi(x, x') \int \frac{dp}{(2\pi)^4} \exp[ip(x - x')] \mathcal{G}(p) \qquad (2.1)$$

where the gauge dependence is isolated in the factor

$$\Phi(x, x') = \exp\left[ieq \int_{x'}^{x} A^{\mu}(\xi) d\xi_{\mu} \right] \qquad (2.2)$$

and where a straight line is taken for the path of integration (and q is the charge matrix). The gauge-independent, momentum space Green's function is[7]

$$\mathcal{G}(p) = i \int_{0}^{\infty} ds \exp\left[-is\left(m^2 + p_{\parallel}^2 + \frac{\tan z}{z} p_{\perp}^2 \right) \right]$$

$$\times \frac{1}{\cos z} \left[(m - \gamma p_{\parallel}) \exp(iq\sigma_3 z) - \frac{1}{\cos z} \gamma p_{\perp} \right] \qquad (2.3)$$

with the notations

$$(ab)_{\parallel} = -a^0 b^0 + a_3 b_3 \qquad (ab)_{\perp} = a_1 b_1 + a_2 b_2 \qquad z = seH \qquad (2.4)$$

the 3-axis being the direction of the magnetic field. This approach is particularly convenient for calculations of processes in which the electrons are all internal, such as vacuum polarization (in both magnetic[12] and parallel electric and magnetic[13] fields), although modifications of the electron propagation function have also been treated[7] in this way. In addition, the following calculations have been performed using (2.3) explicitly: photon decay into neutrino pairs in a strong magnetic field[14] (see Figure 1) and electron pair creation by photon–photon scattering[17] (from forward elastic photon–photon scattering—see Figure 2).

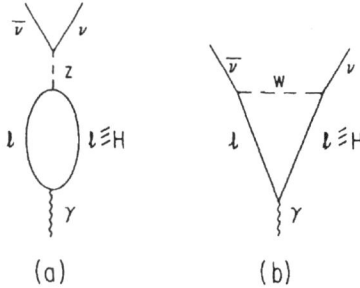

(a)   (b)

FIGURE 1. Lepton exchange processes by which a photon can decay into a neutrino and an antineutrino. In practice, the vector bosons are replaced by point neutral- and charged-current interactions (l = electron or muon, and the $\gtrless$H signifies the presence of an external magnetic field).

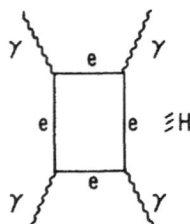

FIGURE 2. From the imaginary part of forward elastic photon–photon scattering, the total cross section for $\gamma\gamma \to e^+e^-$ can be obtained.

However, it is not always most convenient to start with the above closed form for the electron Green's function. This seems to be especially true when the process in question involves external electrons. Thus, for example, one way of calculating synchrotron radiation is to apply the optical theorem to forward Compton scattering,[9,10] described by the vacuum persistence amplitude (see also Figure 3)

$$\langle 0_+|0_-\rangle = -\tfrac{1}{2}i \int dx\, dx'\; \epsilon_\lambda^\mu \psi(x)\gamma^0 M_{\mu\nu}(x, x')\psi(x')\epsilon_\lambda^{\nu*} \qquad (2.5)$$

where the "mass operator" is

$$M_{\mu\nu}(x, x') = -e^2 d\omega_k \gamma_\mu e^{ikx} G_+^A(x, x') e^{-ikx'}\gamma_\nu \qquad (2.6)$$

with the momentum space measure for the photon being

$$d\omega_k = \frac{d\mathbf{k}}{(2\pi)^3}\frac{1}{2k^0} \qquad (2.7)$$

Equation (2.5) is to be evaluated according to the condition that the electron field satisfies the Dirac equation,

$$(m + \gamma\pi)\psi = 0 \qquad (2.8)$$

$$\pi^\mu = (1/i)\partial^\mu - eqA^\mu \qquad (2.9)$$

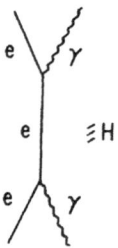

FIGURE 3. One way of computing the power spectrum for synchrotron radiation lies in computing the imaginary part of the second-order forward Compton scattering amplitude.

We write the electron Green's function as

$$G_+(x, x')^A = \langle x| i \int_0^\infty ds\, e^{-is\mathcal{H}}(m - \gamma\pi)|x'\rangle \qquad (2.10)$$

where we have introduced a "Hamiltonian"

$$\mathcal{H} = m^2 - (\gamma\pi)^2 = \pi^2 + m^2 - eq\sigma F \qquad (2.11)$$

The evaluation of the mass operator (2.6) is based upon the recognition[2-4] of $\exp(-is\mathcal{H})$ as a proper-time evolution operator, so for any operator B,

$$B(s) = e^{is\mathcal{H}} B e^{-is\mathcal{H}} \qquad (2.12)$$

which implies the Heisenberg equation of motion

$$\frac{d}{ds} B(s) = (1/i)[B(s), \mathcal{H}] \qquad (2.13)$$

For the operators occurring here, this equation may be easily integrated, yielding the matrix equations

$$\pi(s) = e^{2eqFs}\pi \qquad (2.14a)$$

$$x(s) = x + [(e^{2eqFs} - 1)/eqF]\pi \qquad (2.14b)$$

$$\gamma(s) = e^{2eqFs}\gamma \qquad (2.14c)$$

This, together with the fundamental commutators

$$[\pi, \pi] = ieqF \qquad (2.15a)$$

where F is constant, and

$$[x, \pi] = i1 \qquad (2.15b)$$

makes the commutation of the operators quite straightforward. For further details, see References 9 and 10.

Only slightly more elaborate is the situation when the intermediate state is a two-particle system. An example of this occurs in the calculation of the cross section for Compton scattering in a homogeneous magnetic field which we have computed for the situation when the incident photon direction is parallel to the magnetic field—see Figure 4.[15,16] In such a case one combines the propagation functions of photon and electron according to

$$\frac{1}{k^2} \frac{1}{(\pi - k)^2 + m^2 - eq\sigma F} = -\int_0^\infty ds\, s \int_0^1 du\, e^{-is\mathcal{H}(u)} \qquad (2.16)$$

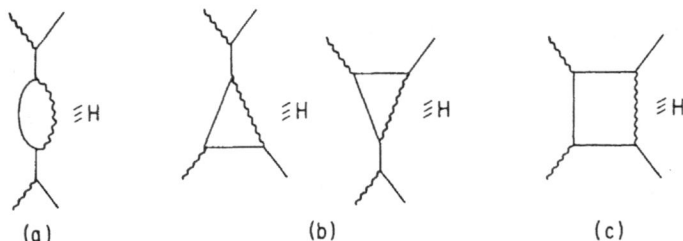

FIGURE 4. The total cross section for Compton scattering in an external field can be obtained from the imaginary part of the fourth-order forward Compton scattering amplitudes illustrated here.

where the Hamiltonian is now

$$\mathscr{H}(u) = (k - u\pi)^2 + u(1 - u)\pi^2 + u(m^2 - eq\sigma F) \qquad (2.17)$$

Again, the operator commutations may be carried out in a straightforward manner by regarding $\exp[-is\mathscr{H}(u)]$ as a time evolution operator. Here, however, it is convenient to define[3] an operator ξ_μ complementary to the photon momentum k_μ,

$$[\xi_\mu, k_\mu] = ig_{\mu\nu} \qquad (2.18)$$

leading to the following set of time-evolved operator equations:

$$k(s) = k \qquad (2.19\text{a})$$

$$eqF[\xi(s) - \xi] = [e^{2ueqFs} - 1 + 2(1 - u)eqFs]k$$
$$- (e^{2ueqFs} - 1)\pi \qquad (2.19\text{b})$$

$$\pi(s) - k = e^{2ueqFs}(\pi - k) \qquad (2.19\text{c})$$

$$\gamma(s) = e^{2ueqFs}\gamma \qquad (2.19\text{d})$$

By introducing ξ, integration over k is equivalent to taking an expectation value in an eigenstate of ξ,

$$\int \frac{dk}{(2\pi)^4} f(k) = \langle \xi' = 0 | f(k) | \xi' = 0 \rangle \qquad (2.20)$$

For further details of this calculation see References 15 and 16. Calculations making use of this proper-time approach also include that of Adler on photon splitting and dispersion,[24] as well as computations of synchrotron radiation,[3-6,9,10] and of photon pair creation and attentuation.[11]

To summarize, both approaches to strong-field problems discussed in this section make use of, in an essential way, proper-time representations. In the first, the closed form, the proper-time electron Green's function is employed, while in the second we solve the proper-time equations of motion for the operators to perform the commutations. (This is the way the electron Green's function was first worked out, in fact.[2]) These methods directly supply total rates, such as total cross sections or total power radiated, but differential quantities, such as the power spectrum for synchrotron radiation, can be obtained with little more trouble. These methods are to be contrasted with conventional approaches, which make use of explicit particle wave functions,[25] which are more cumbersome, less elegant to use, and usually contain more information than is required. The proper-time methods, it need hardly be said, have many other applications than to strong-field problems—an example is the efficient calculation of the anomalous magnetic moment of the electron.[2,26]

3. Casimir Effect

Now we turn to a complementary situation, one in which it is the photon Green's function that is modified by the medium. The Casimir force between macroscopic dielectric or conducting bodies arises because that Green's function changes when the position of the bodies is altered. This in turn implies a change in the energy of that system and, by the principle of virtual work, a force between the bodies.

A convenient starting point for discussing the Casimir effect in dielectrics is the action expression[22]

$$W = \int dx \, [\mathbf{P} \cdot (-\mathbf{A} - \nabla\phi) + \epsilon \mathbf{E} \cdot (-\mathbf{A} - \nabla\phi) \\ - \mathbf{H} \cdot (\nabla \times \mathbf{A}) + \tfrac{1}{2}H^2 - \tfrac{1}{2}\epsilon E^2] \quad (3.1)$$

where $\mathbf{P}$ is an external polarization source. Variation with respect to $\mathbf{A}$, ϕ, $\mathbf{E}$, and $\mathbf{H}$ supplies the four Maxwell equations. We define a Green's dyadic Γ, which relates $\mathbf{E}$ to its source $\mathbf{P}$,

$$\mathbf{E}(x) = \int dx' \, \Gamma(x, x') \cdot \mathbf{P}(x') \quad (3.2)$$

in terms of which W is

$$W = \tfrac{1}{2} \int dx \, dx' \, \mathbf{P}(x) \cdot \Gamma(x, x') \cdot \mathbf{P}(x') \quad (3.3)$$

Now, if the dielectrics are moved infinitesimally, there is an infinitesimal alteration in the dielectric constant, $\delta\epsilon$, which implies, by the principle of stationary action, that

$$\delta_\epsilon W = \int dx \, \delta\epsilon \tfrac{1}{2}E^2 \quad (3.4)$$

Such a virtual motion gives rise to an effective polarization source, the numerical measure of which can be obtained by comparing (3.4) with

(3.1):

$$iP(x)P(x')|_{\text{eff}} = 1\delta\epsilon\,\delta(x - x') \tag{3.5}$$

When this is inserted into (3.3), we obtain the change in the action in terms of the Green's dyadic,

$$\delta W = -\int dt\,\delta E \tag{3.6a}$$

where

$$\delta E = \frac{i}{2}\int d\mathbf{r}\,\frac{d\omega}{2\pi}\,\delta\epsilon(\mathbf{r}, \omega)\,\text{Tr}\,\mathbf{\Gamma}(\mathbf{r}, \mathbf{r}, \omega) \tag{3.6b}$$

is the change in the energy of the system. We have in the last step incorporated dispersion. (Temperature effects may be included in a standard way by using instead a Fourier series representation in imaginary time: formally in the above we make the following replacements:

$$\int\frac{d\omega}{2\pi} \to \frac{i}{\beta}\sum_{n=-\infty}^{\infty} \quad \text{and} \quad \omega \to \frac{2\pi i n}{\beta} \tag{3.7}$$

where $\beta = 1/kT$.) The determination of the energy change and hence the force between the bodies thus reduces to finding the Green's dyadic $\mathbf{\Gamma}$ subject to the boundary conditions at the interfaces between the dielectrics, that $\mathbf{H}$, $\mathbf{E}_\parallel$, and $\epsilon\mathbf{E}_\perp$ be continuous. The magnetic field corresponds to the curl of $\mathbf{\Gamma}$, just as the electric field is directly expressible in terms of $\mathbf{\Gamma}$:

$$i\mathbf{E}(x)\mathbf{E}(x')|_{\text{eff}} = \int_{-\infty}^{\infty}\frac{d\omega}{2\pi}\,e^{-i\omega(t-t')}\mathbf{\Gamma}(\mathbf{r}, \mathbf{r}', \omega) \tag{3.8a}$$

$$i\mathbf{H}(x)\mathbf{H}(x')|_{\text{eff}} = -\int_{-\infty}^{\infty}\frac{d\omega}{2\pi}\,e^{-i\omega(t-t')}\frac{1}{\omega^2}\,\boldsymbol{\nabla}\times\mathbf{\Gamma}(\mathbf{r}, \mathbf{r}', \omega)\times\boldsymbol{\nabla}' \tag{3.8b}$$

The Green's function equation satisfied by $\mathbf{\Gamma}$ is seen from (3.1) to be

$$-\boldsymbol{\nabla}\times(\boldsymbol{\nabla}\times\mathbf{\Gamma}) + \omega^2\epsilon\mathbf{\Gamma} = -\omega^2\mathbf{1}\delta(\mathbf{r}, \mathbf{r}') \tag{3.9}$$

So the Casimir problem, which is essentially quantum mechanical, reduces to the solution of a classical boundary-value problem. The examples so far treated include the calculation of the force between semi-infinite, parallel dielectrics, separated by a third dielectric.[22,27] The case of perfect parallel conducting plates separated by vacuum[19,21,28] can be obtained, roughly, by the limit $\epsilon \to \infty$. By taking the opposite limit, $\epsilon \to 0$, corresponding to tenuous media, the long- and short-range van der Waals forces can be inferred.[20] Of particular interest to condensed matter physics is a suggestive but incomplete calculation, based on these ideas, of the Casimir contribution to the surface tension of an ideal liquid (liquid He?). The result is strongly cutoff dependent, signaling a breakdown of the macroscopic approach, but with a reasonable value for the cutoff parameter, both the surface tension and the latent heat of liquid helium are reproduced to within a factor of two or three.[22] An amusing calculation of (presumably) only theoretical interest is that of the Casimir energy of a perfectly conducting spherical shell.[23] In contradiction to Casimir's hopes[29] to obtain in this way a semiclassical model of the electron, the self-force turns out to be repulsive, as first found by Boyer.[30]

The simplicity and power of this approach to Casimir forces should be noted. (A variant method makes use of the photon stress tensor.[22,23]) The basis of most other calculations[28] of this phenomenon is the zero-point energy of the vacuum which, of course, is infinite. Besides the renormalization difficulties, such calculations employ complicated mode summations, which are obviated in our approach, which makes use of a closed form for the (reduced) photon Green's function.

4. Conclusions

Because it deals directly with physical states and not with operator-valued fields, source theory is very effective at bridging the gap between classical and quantum field theory. An example is given by the very simple, physical derivations of the observable consequences of general relativity.[31] In this review we have discussed two categories of semi-classical processes—effects due to the motion of electrons in prescribed fields and effects due to the modification of the (virtual) propagation of photons by material media (the Casimir effect). In both we consider the quantum behavior of particles in a prescribed macroscopic, classical environment.

It is evident that not only is there more work to be done in the areas surveyed here, but that similar approaches would bring success in other areas beyond that of particle physics. In particular, application of source theory ideas to statistical physics and many-body physics is overdue. Perhaps this article will stimulate that development.

References and Notes

1. J. Schwinger, *Particles, Sources, and Fields,* Vols. I and II, Addison-Wesley, Reading, Massachusetts (1970 and 1973).

2. J. Schwinger, *Phys. Rev.* **82**, 664 (1951).
3. J. Schwinger, *Phys. Rev. D* **7**, 1696 (1973).
4. J. Schwinger, *Particles, Sources, and Fields,* Vol. III, Section 5-6, in preparation.
5. W.-y. Tsai and A. Yildiz, *Phys. Rev. D* **8**, 3446 (1973).
6. W.-y. Tsai, *Phys. Rev. D* **8**, 3460 (1973).
7. W.-y. Tsai, *Phys. Rev. D* **10**, 1342 (1974).
8. J. Schwinger and W.-y. Tsai, *Phys. Rev. D* **9**, 1843 (1974).
9. J. Schwinger and W.-y. Tsai, *Ann. Phys. (N.Y.)* **110**, 63 (1978).
10. W.-y. Tsai, *Phys. Rev. D* **18**, 3863 (1978).
11. W.-y. Tsai and T. Erber, *Phys. Rev. D* **10**, 492 (1974) and **12**, 1132 (1975).
12. W.-y. Tsai, *Phys. Rev. D* **10**, 2699 (1974).
13. L. F. Urrutia, *Phys. Rev. D* **17**, 1977 (1978).
14. L. L. DeRaad, Jr., K. A. Milton, and N. D. Hari Dass, *Phys. Rev. D* **14**, 3326 (1976).
15. L. L. DeRaad, Jr., N. D. Hari Dass, and K. A. Milton, *Phys. Rev. D* **9**, 1041 (1974).
16. K. A. Milton, W.-y. Tsai, L. L. DeRaad, Jr., and N. D. Hari Dass, *Phys. Rev. D* **10**, 1299 (1974).
17. Y. J. Ng and W.-y. Tsai, *Phys. Rev. D* **16**, 286 (1977).
18. J. Schwinger, W.-y. Tsai, and T. Erber, *Ann. Phys. (N.Y.)* **96**, 303 (1976); T. Erber, D. White, W.-y. Tsai, and H. G. Latal, *Ann. Phys. (N.Y.)* **102**, 405 (1976).
19. H. B. G. Casimir, *Proc. K. Ned. Akad. Wet. Ser. B* **51**, 793 (1948).
20. H. B. G. Casimir and D. Polder, *Phys. Rev.* **73**, 360 (1948).
21. J. Schwinger, *Lett. Math. Phys.* **1**, 43 (1975).
22. J. Schwinger, L. L. DeRaad, Jr., and K. A. Milton, *Ann. Phys. (N.Y.)* **115**, 1 (1978).
23. K. A. Milton, L. L. DeRaad, Jr., and J. Schwinger, *Ann. Phys. (N.Y.)* **115**, 388 (1978).
24. S. L. Adler, *Ann. Phys. (N.Y.)* **67**, 599 (1971).
25. See, for example, I. M. Ternov, V. G. Bagrov, V. A. Bordovttsyn, and O. F. Dorofeev, *Zh. Eksp. Teor. Fiz.* **55**, 2273 (1968) [English translation: *Sov. Phys. JETP* **28**, 1206 (1969)].
26. J. Schwinger, *Particles, Sources, and Fields,* Vol. III, Section 5-7, in preparation; K. A. Milton, W.-y. Tsai, and L. L. DeRaad, Jr., *Phys. Rev. D* **9**, 1809, 1814 (1974).
27. The result agrees with that obtained first by E. M. Lifshitz, *Zh. Eksp. Teor. Fiz.* **29**, 94 (1955) [English translation: *Sov. Phys. JETP* **2**, 73 (1956)], and confirmed experimentally at $T = 0$ most convincingly by E. S. Sabisky and C. H. Andersen, *Phys. Rev. A* **7**, 790 (1973).
28. The earlier literature in this nearly vast subject is surveyed by T. H. Boyer, *Ann. Phys. (N.Y.)* **56**, 474 (1970).
29. H. B. G. Casimir, *Physica* **19**, 846 (1956).
30. T. H. Boyer, *Phys. Rev.* **174**, 1764 (1968).
31. Reference 1, Vol. I, pp. 78-85; J. Schwinger, *Am. J. Phys.* **42**, 507, 510 (1974); K. A. Milton, *Am. J. Phys.* **42**, 911 (1974).

16

A Local Gauge-Invariant Formulation of Quantum Electrodynamics

D. H. Sharp

1. Introduction

In this article I want to discuss a local and manifestly gauge-invariant formulation of quantum electrodynamics which has been developed by R. Menikoff and myself.[1,2] The manifest gauge invariance is achieved by using the electromagnetic field strengths, rather than potentials, to describe the electromagnetic field, and by using local currents, rather than canonical fields, to describe the matter. In order to represent this theory in Hilbert space, we study the continuous unitary representations of the group obtained from the exponentiated currents and electromagnetic field strengths. These operators can be represented on a Hilbert space having positive norm, so that the necessity for an indefinite metric does not arise here, and the equations of motion hold as operator equations in this Hilbert space.

From the mathematical point of view, a motivation for studying a manifestly gauge-invariant formulation of quantum electrodynamics derives from the rather complicated state of affairs which arises when one employs the usual potentials in quantum field theory.[3] For example, one must require the physical observables $\mathbf{E}(\mathbf{x})$, $\mathbf{B}(\mathbf{x})$ which act on the physical Hilbert space, to be the same in different gauges, but one cannot in general represent the vector potentials in two different gauges on the same Hilbert space. From a more practical standpoint, it might be useful to have a perturbation theory in which each individual diagram is gauge

D. H. Sharp • Theoretical Division, Los Alamos Scientific Laboratory, University of California, Los Alamos, New Mexico 87545. Work supported by the U.S. Department of Energy.

invariant, as would presumably result from a small-coupling analysis of the formulation of quantum electrodynamics to be discussed here.

Manifestly gauge-invariant formulations of quantum electrodynamics have been developed previously by a number of authors. For example, DeWitt[4] and Mandelstam[5] showed how quantum electrodynamics can be written in a formally gauge-invariant way by introducing path-dependent fields. These, however, result in a nonlocal quantum field theory.[3]

An alternative approach was outlined by the present author who showed,[1] again on a formal level, that the electrodynamics of relativistic charged scalar mesons could be written in a *local* and manifestly gauge-invariant way if the mesons were described using local currents. In exploring this idea further in Reference 2, we considered nonrelativistic particles interacting with an electromagnetic field, because in this case a good deal more is known about the mathematical properties of local currents.

Although the formalism to be reviewed here can be extended to include models with magnetic monopoles or several species of charged particles, for simplicity we shall restrict our discussion to the case of a single species of particle with an electric charge. Sections 2 and 3 contain a review of our principal results, while in Section 4 we call attention to a number of interesting questions raised by this work.

2. Quantum Electrodynamics Using Local Currents

We first recall that a completely consistent quantization of the *free* electromagnetic field in terms of $\mathbf{E}(\mathbf{x})$ and $\mathbf{B}(\mathbf{x})$ was given long-ago by Jordan and Pauli.[6] This work is reviewed and formulated in terms of the representation theory which will be used in this paper in Reference 2. Also, there have been several studies[7,8] of the formulation of nonrelativistic quantum theory in terms of local currents in the case when electromagnetic interactions are absent. Consequently, we will proceed directly to the case of interest here—nonrelativistic quantum electrodynamics.

2.1. First Formulation: Algebra of the Operators $\rho(\mathbf{x})$, $\mathbf{J}(\mathbf{x})$, $\mathbf{E}(\mathbf{x})$, and $\mathbf{B}(\mathbf{x})$

We use the canonical fields as a heuristic to motivate the form of the local current algebra and the Hamiltonian. The number density of particles $\rho(\mathbf{x})$, the flux density of particles $\mathbf{J}(\mathbf{x})$, and the electric and magnetic fields are given by

$$\rho(\mathbf{x}) = \psi^\dagger(\mathbf{x})\psi(\mathbf{x})$$

$$\mathbf{J}(\mathbf{x}) = \frac{1}{2im}\left[\psi^\dagger(\mathbf{x})\left(\hbar\nabla - \frac{ie}{c}\mathbf{A}(\mathbf{x})\right)\psi(\mathbf{x}) - \left(\hbar\nabla + \frac{ie}{c}\mathbf{A}(\mathbf{x})\right)\psi^\dagger(\mathbf{x})\psi(\mathbf{x})\right]$$

$$\mathbf{E}(\mathbf{x}) = -\nabla A_0(\mathbf{x}) - \frac{1}{c}\frac{\partial \mathbf{A}}{\partial t}$$

$$\mathbf{B}(\mathbf{x}) = (\nabla \times \mathbf{A})(\mathbf{x})$$

These quantities are all physical observables and hence gauge invariant. Note that the mass density is $m\rho(\mathbf{x})$, the charge density is $e\rho(\mathbf{x})$, the electric current density is $e\mathbf{J}(\mathbf{x})$, and the particles' momentum density is $m\mathbf{J}(\mathbf{x})$. From the commutation relations for the canonical fields

$$[\psi(\mathbf{x}), \psi^\dagger(\mathbf{y})]_\pm = \delta(\mathbf{x} - \mathbf{y})$$

$$[A_\mu(\mathbf{x}), \mathring{A}_\nu(\mathbf{y})] = -c^2 i\hbar g_{\mu\nu}\delta(\mathbf{x} - \mathbf{y})$$

one obtains the following equal-time commutation relations among the operators ρ, $\mathbf{J}$, $\mathbf{E}$, and $\mathbf{B}$:

$$[\rho(\mathbf{x}), J_i(\mathbf{y})] = -i\left(\frac{\hbar}{m}\right)\frac{\partial}{\partial x_i}[\delta(\mathbf{x} - \mathbf{y})\rho(\mathbf{x})] \tag{2.1}$$

$$[J_i(\mathbf{x}), J_k(\mathbf{y})] = i\left(\frac{\hbar}{m}\right)\frac{\partial}{\partial y_i}[\delta(\mathbf{x} - \mathbf{y})J_k(\mathbf{y})] - i\left(\frac{\hbar}{m}\right)\frac{\partial}{\partial x_k}[\delta(\mathbf{x} - \mathbf{y})J_i(\mathbf{x})]$$

$$+ i\left(\frac{\hbar e}{m^2 c}\right)\epsilon_{ikj}\rho(\mathbf{x})B_j(\mathbf{x})\delta(\mathbf{x} - \mathbf{y}) \tag{2.2}$$

$$[E_i(\mathbf{x}), B_k(\mathbf{y})] = i(c\hbar)\epsilon_{ikj}\frac{\partial}{\partial y_i}\delta(\mathbf{x} - \mathbf{y}) \tag{2.3}$$

$$[J_i(\mathbf{x}), E_k(\mathbf{y})] = i\left(\frac{\hbar e}{m}\right)\rho(\mathbf{x})\delta_{ik}\delta(\mathbf{x} - \mathbf{y}) \tag{2.4}$$

with all other commutators vanishing. Thus, under equal-time commutation the local currents and the components of the electromagnetic field form a closed algebra. However, the algebra is nonlinear owing to the $\rho(\mathbf{x})\mathbf{B}(\mathbf{x})$ term which appears in equation (2.2), and hence it is not a Lie algebra. Another important property of the algebra is that the interaction and the coupling constant enter explicitly in the commutation relations (2.2) and (2.4). As a result, the algebra (2.1)–(2.4) for the interacting theory is different from that of the free theory and hence a different representation of the local currents and fields is required. Furthermore, different representations are needed for different values of the coupling constant.

We next consider the Hamiltonian. In terms of the canonical fields it is given by

$$H = \frac{\hbar^2}{2m} \int d^3x \left(\boldsymbol{\nabla} + i\frac{e}{\hbar c}\mathbf{A}(\mathbf{x}) \right) \psi^\dagger(\mathbf{x}) \left(\boldsymbol{\nabla} - i\frac{e}{\hbar c}\mathbf{A}(\mathbf{x}) \right) \psi(\mathbf{x})$$
$$+ \tfrac{1}{2} \int d^3x \, (\mathbf{E}^2 + \mathbf{B}^2)(\mathbf{x})$$

By using the local currents, this can be written as

$$H = \frac{\hbar^2}{8m} \int d^3x \, K_i^\dagger(\mathbf{x}) \frac{1}{\rho(\mathbf{x})} K_i(\mathbf{x}) + \tfrac{1}{2} \int d^3x \, (E^2 + B^2)(\mathbf{x}) \qquad (2.5)$$

where

$$\mathbf{K}(\mathbf{x}) = \boldsymbol{\nabla}\rho(\mathbf{x}) + (2im/\hbar)\mathbf{J}(\mathbf{x}) \qquad (2.5')$$

At this point, one can forget about the canonical fields and take the local currents, their commutation relations [equations (2.1)–(2.4)], and equation (2.5) for the Hamiltonian as defining the theory.

It should be noted that the expression "$1/\rho(\mathbf{x})$" occurring in equation (2.5) cannot be interpreted as a Hilbert space operator. However, in the case when electromagnetic interactions are absent, it has been shown[9,10] that two properties of the theory prevent this from leading to an ill-defined expression for the Hamiltonian density $H(\mathbf{x})$. First, $1/\rho(\mathbf{x})$ always occurs sandwiched between $\mathbf{K}(\mathbf{x})$ and $\mathbf{K}^\dagger(\mathbf{x})$, so that when matrix elements of the Hamiltonian are taken, $1/\rho(\mathbf{x})$ does not act on a Hilbert space vector Φ, but on the *vector-valued distribution* $\mathbf{K}(\mathbf{x})\Phi$. Second, $\mathbf{K}(\mathbf{x})\Phi$ turns out to be "proportional" to $\rho(\mathbf{x})$, in a certain technical sense. As a result, $1/\rho(\mathbf{x})$ can be given a direct mathematical definition in such a way that $H(f) = \int H(\mathbf{x})f(x) \, d\mathbf{x}$ is actually a well-defined operator in Hilbert space. It seems very likely that this approach can be extended to the Hamiltonian (2.5) as well.

As one general check on the consistency of this formulation, it can be verified[2] that equations (2.1)–(2.5) lead to the correct operator form of the classical equations which determine the time development of $\rho(\mathbf{x}, t)$, $\mathbf{J}(\mathbf{x},t)$, $\mathbf{E}(\mathbf{x},t)$, and $\mathbf{B}(\mathbf{x}, t)$. The two Maxwell equations $\boldsymbol{\nabla}\cdot\mathbf{E} = e\rho$ and $\boldsymbol{\nabla}\cdot\mathbf{B} = 0$, however, must be imposed as initial-value equations in order for the total momentum operator to be identifiable as the generator of translations in space. This point is discussed in detail in Section 2.2.

An interesting aspect of this formulation is that the Hamiltonian (2.5) is, formally, the sum of the free Hamiltonian for the particles plus the free Hamiltonian for the electric and magnetic fields. The only place where the interaction appears explicitly is in the equal-time algebra (2.1)–(2.4).

2.2. Second Formulation: Algebra of the Operators $\rho(\mathbf{x})$, $\mathbf{P}(\mathbf{x})$, $\mathbf{E}(\mathbf{x})$, and $\mathbf{B}(\mathbf{x})$

To discuss the representation theory of the local currents, it is convenient to make a change of variables so as to obtain a (nuclear) Lie algebra. For this purpose we introduce the total momentum density:

$$\mathbf{P}(\mathbf{x}) = m\mathbf{J}(\mathbf{x}) + (1/2c)[\mathbf{E}(\mathbf{x}) \times \mathbf{B}(\mathbf{x}) - \mathbf{B}(\mathbf{x}) \times \mathbf{E}(\mathbf{x})] \qquad (2.6)$$

Next we determine the equal-time algebra generated by $\rho(\mathbf{x})$, $\mathbf{P}(\mathbf{x})$, $\mathbf{E}(\mathbf{x})$, and $\mathbf{B}(\mathbf{x})$. A straightforward calculation shows that the commutation relations involving $\mathbf{P}(\mathbf{x})$ are given by[2]

$$[\rho(\mathbf{x}), P_i(\mathbf{y})] = -i\hbar \frac{\partial}{\partial x_i}[\delta(\mathbf{x}-\mathbf{y})\rho(\mathbf{x})] \qquad (2.7)$$

$$[P_i(\mathbf{x}), P_k(\mathbf{y})] = i\hbar \frac{\partial}{\partial y_i}[\delta(\mathbf{x}-\mathbf{y})P_i(\mathbf{y})] - i\hbar \frac{\partial}{\partial x_k}[\delta(\mathbf{x}-\mathbf{y})P_i(\mathbf{x})]$$
$$+ i(\hbar/c)\epsilon_{ikj}\{[\boldsymbol{\nabla}\cdot\mathbf{E}(\mathbf{x}) - e\rho(\mathbf{x})]B_j(\mathbf{x}) + \boldsymbol{\nabla}\cdot\mathbf{B}(\mathbf{x})E_j(\mathbf{x})\}$$
$$\times \delta(\mathbf{x}-\mathbf{y}) \qquad (2.8)$$

$$[E_i(\mathbf{x}), P_k(\mathbf{y})] = -i\hbar \frac{\partial}{\partial x_k}[E_i(\mathbf{x})\delta(\mathbf{x}-\mathbf{y})] - i\hbar\delta_{ik}\frac{\partial}{\partial y_j}[E_j(\mathbf{y})\delta(\mathbf{x}-\mathbf{y})]$$
$$+ i\hbar\delta_{ik}[\boldsymbol{\nabla}\cdot\mathbf{E}(\mathbf{x}) - e\rho(\mathbf{x})]\delta(\mathbf{x}-\mathbf{y}) \qquad (2.9)$$

$$[B_i(\mathbf{x}), P_k(\mathbf{y})] = -i\hbar \frac{\partial}{\partial x_k}[B_i(\mathbf{x})\delta(\mathbf{x}-\mathbf{y})] - i\hbar\delta_{ik}\frac{\partial}{\partial y_j}[B_j(\mathbf{y})\delta(\mathbf{x}-\mathbf{y})]$$
$$+ i\hbar\delta_{ik}\boldsymbol{\nabla}\cdot\mathbf{B}(\mathbf{x})\delta(\mathbf{x}-\mathbf{y}) \qquad (2.10)$$

For a consistent physical interpretation, one must require

$$\mathbf{P} = \int d^3x\, \mathbf{P}(\mathbf{x})$$

to be the total momentum, i.e., the generator of space translations. The commutator of P with any local operator $\mathcal{O}(\mathbf{x})$ must therefore have the form

$$[\mathcal{O}(\mathbf{x}), P] = -i\hbar\boldsymbol{\nabla}\mathcal{O}(\mathbf{x}) \qquad (2.11)$$

It is easily seen that, *in order for equation (2.11) to be consistent with the above commutation relations,* it is necesssary and sufficient to impose the constraints

$$(\nabla\cdot\mathbf{E})(\mathbf{x}) = e\rho(\mathbf{x}) \tag{2.12}$$

$$(\nabla\cdot\mathbf{B})(\mathbf{x}) = 0 \tag{2.13}$$

These constraints can be interpreted as initial-value equations since it can be shown that if equations (2.12) and (2.13) hold at a fixed time t, then they hold for all time, as a consequence of the equations of motion. The connection between the initial-value equations and the gauge invariance of quantum electrodynamics when formulated using potentials will be discussed in Section 2.3.

Thus we see that, with the imposition of the initial-value equations, the operators $\rho(\mathbf{x})$, $\mathbf{P}(\mathbf{x})$, $\mathbf{E}(\mathbf{x})$, and $\mathbf{B}(\mathbf{x})$ form a local Lie algebra defined by the commutation relations (2.3) and (2.7)–(2.10), with all other commutators vanishing.

We can also describe the dynamics in terms of these variables. For example, the Hamiltonian is still given by equation (2.5), but with $\mathbf{K}(\mathbf{x})$ given by

$$\mathbf{K}(\mathbf{x}) = \nabla\rho(\mathbf{x}) + (2i/\hbar)[\mathbf{P}(\mathbf{x}) - (1/2c)(\mathbf{E}\times\mathbf{B} - \mathbf{B}\times\mathbf{E})(\mathbf{x})] \tag{2.14}$$

instead of (2.5′). The interaction now appears in the Hamiltonian, as usual, instead of in the algebra. Note that the coupling constant e appears only in the initial-value equation (2.12). If we choose the charge density $e\rho(\mathbf{x})$ as our variable instead of the number density $\rho(\mathbf{x})$, then the coupling constant would appear only in the Hamiltonian.

2.3. Gauge Invariance

At this point we discuss the relationship between the formulation of electrodynamics using local currents and that using potentials, particularly with regard to the role of gauge invariance. In either case one obtains Maxwell's equations, but they come about in different ways.

In terms of the $\mathbf{E}$ and $\mathbf{B}$ fields, the dynamics is determined by two first-order equations of motion:

$$(1/c)\partial\mathbf{B}(\mathbf{x})/\partial t = -(\nabla\times\mathbf{E})(\mathbf{x}) \tag{2.15}$$

$$(1/c)\partial\mathbf{E}(\mathbf{x})/\partial t = -(e/c)\mathbf{J}(\mathbf{x}) + (\nabla\times\mathbf{B})(\mathbf{x}) \tag{2.16}$$

In addition two initial-value equations, (2.12) and (2.13), must be imposed in order for the theory to accommodate a representation of the translation group. Together, equations (2.15), (2.16) and (2.12), (2.13) comprise Maxwell's equations.

Alternatively, one can write Maxwell's equations in the form

$$\partial_\mu F^{\mu\nu} = j^\nu \tag{2.17}$$

$$\partial_\mu {}^*F^{\mu\nu} = 0 \tag{2.18}$$

where $^*F^{\mu\nu}$ is the dual of $F^{\mu\nu}$. Equation (2.18) will be satisfied identically if $F^{\mu\nu}$ is written as the curl of a four-vector potential A^μ; $F^{\mu\nu} = \partial^\mu A^\nu - \partial^\nu A^\mu$. However, the transformation from $F^{\mu\nu}$ to A^μ is not unique. To determine A^μ uniquely, a constraint, or gauge condition, must be imposed. The equation of motion (second order in time for A) is then equivalent to equation (2.17). Thus, equation (2.17) is the dynamical equation when the theory is expressed in terms of potentials.

3. Representation Theory

Many mathematicical properties of this model can be conveniently analyzed by studying the representations of the group obtained from equations (2.3) and (2.7)–(2.10) by exponentiating the currents and fields. Defining

$$\mathcal{V}(\phi_t^g) = \exp[itP(\mathbf{g})]$$

where $P(\mathbf{g}) = \int d^3x\, \mathbf{P}(\mathbf{x})\cdot\mathbf{g}(\mathbf{x})$ and ϕ_t^g is the flow, associated with the vector field $\mathbf{g}(\mathbf{x})$, defined by the equation

$$\frac{\partial \phi_t^g(\mathbf{x})}{\partial t} = \mathbf{g}(\phi_t^g(\mathbf{x}))$$

one can write an arbitrary group element as

$$\Gamma(f, \mathbf{g}, \mathbf{h}, \phi) = \exp[i\rho\,(f)]\exp[iE(\mathbf{g})]\exp[iB(\mathbf{h})]\mathcal{V}(\phi) \tag{3.1}$$

and show that the law for multiplying two such elements together is given by[2]

$$\Gamma(f_1, \mathbf{g}_1, \mathbf{h}_1, \phi_1)\Gamma(f_2, \mathbf{g}_2, \mathbf{h}_2, \phi_2)$$
$$= \exp\left[i\int d^3x\, \mathbf{W}_{\phi_1}(\mathbf{g}_2)(\mathbf{x})\cdot(\mathbf{\nabla}\times\mathbf{h}_1)(\mathbf{x})\right]$$
$$\times \Gamma(f_1 + f_2 \circ \phi_1, \mathbf{g}_1 + \mathbf{W}_{\phi_1}(\mathbf{g}_2), \mathbf{h}_1 + \mathbf{W}_{\phi_1}(\mathbf{h}_2), \phi_2 \circ \phi_1) \tag{3.2}$$

In equation (3.2), "$\circ$" stands for the operation of composition of functions and $\mathbf{W}_\phi(\mathbf{g})(\mathbf{x}) = (\mathbf{\nabla}\phi_k)(\mathbf{x})(g_k \circ \phi)(\mathbf{x})$.

We can describe the representations of this group in terms of a measure and a "system of multipliers" using the formalism developed by Gel'fand and Vilenkin.[11] Here we shall simply summarize the results of this theory as they apply in the present case. (See References 10 and 12 for proofs in the case of the ρ, $\mathbf{J}$ current algebra.)

We begin by considering the generating functional

$$L(\mathbf{g}) = (\Omega, \exp[iE(\mathbf{g})]\Omega) \tag{3.3}$$

where Ω is a cyclic vector for the representation, which we will always identify with the ground state of the Hamiltonian [equations (2.5) and (2.14)]. One can write $L(\mathbf{g})$ as the Fourier transform of a positive measure μ on $\mathscr{S}'$, the real continuous dual of the space of test functions. Thus,

$$L(\mathbf{g}) = \int_{\mathscr{S}'} d\mu(\mathbf{G})\exp[i(\mathbf{G}, \mathbf{g})] \tag{3.4}$$

where $(\mathbf{G}, \mathbf{g})$ means the functional $\mathbf{G}$ evaluated at $\mathbf{g}(\mathbf{x})$. The measure μ can be used to define a Hilbert space $\mathscr{H} = L^2_\mu(\mathscr{S}')$. The group elements act in the following way on this Hilbert space. Let $\Psi(\mathbf{G}) \in \mathscr{H}$. Then:

(1) The element $\exp[iE(\mathbf{g})]$ is represented as multiplication by $\exp[i(\mathbf{G}, \mathbf{g})]$, i.e.,

$$(\exp[iE(\mathbf{g})]\Psi)(\mathbf{G}) = \exp[i(\mathbf{G}, \mathbf{g})]\Psi(\mathbf{G}) \tag{3.5}$$

(2) The operator $\rho(f)$ is defined by equation (2.12), or $e\rho(f) = -E(\nabla f)$. As a result

$$(\exp[ie\rho(f)]\Psi)(\mathbf{G}) = \exp[-i(\mathbf{G}, \nabla f)]\Psi(\mathbf{G}) \tag{3.6}$$

(3) Next we have

$$(\mathscr{V}(\boldsymbol{\phi})\Psi)(\mathbf{G}) = \chi_\phi(\mathbf{G})\Psi(\phi^*\mathbf{G})[d\mu(\phi^*\mathbf{G})/d\mu(\mathbf{G})]^{1/2} \tag{3.7}$$

where ϕ^* is a map from $\mathscr{S}'$ into $\mathscr{S}'$ defined by

$$(\phi^*\mathbf{G}, \mathbf{g}) = (\mathbf{G}, \bar{\phi}\mathbf{g}) \tag{3.8}$$

with

$$(\bar{\phi}\mathbf{g})_j(\mathbf{x}) = (\partial_j \phi_k)(\mathbf{x})(g_k \circ \boldsymbol{\phi})(\mathbf{x})$$

Also, $\chi_\phi(\mathbf{G})$ is a multiplier for $\mathscr{V}(\boldsymbol{\phi})$, and $d\mu(\phi^*\mathbf{G})/d\mu(\mathbf{G})$ is the Radon–Nikodym derivative. For this derivative to exist, it is necessary that the measure μ be quasi-invariant.

(4) Finally,

$$(\exp[iB(\mathbf{h})]\Psi)(\mathbf{G}) = Z_h(\mathbf{G})\Psi(\mathbf{h}^*\mathbf{G})[d\mu(\mathbf{h}^*\mathbf{G})/d\mu(\mathbf{G})]^{1/2} \tag{3.9}$$

where $\mathbf{h}^*$ is another map from $\mathscr{S}'$ into $\mathscr{S}'$, here defined by

$$(\mathbf{h}^*\mathbf{G}, \mathbf{g}) = (\mathbf{G} + \nabla \times \mathbf{h}, \mathbf{g}) \tag{3.10}$$

and $Z_h(\mathbf{G})$ is a multiplier for $B(\mathbf{h})$.

The multipliers χ_ϕ and Z_h are complex-valued functions of modulus one. For the group multiplication law (3.2) to be obeyed, the multipliers must satisfy

$$\chi_{\phi_1}(G)\chi_{\phi_2}(\phi_1^*G) = \chi_{\phi_2 \circ \phi_1}(G)$$

$$Z_{h_1}(G)Z_{h_2}(h_1^*G) = Z_{h_1+h_2}(G) \qquad (3.11)$$

$$\chi_\phi(G)Z_h(\phi^*G) = Z_{\dot\phi h}(G)\chi_\phi((\dot\phi h)^*G)$$

While equations (3.11) follow from the general representation theory, an additional constraint on the multipliers follows from the initial-value equation $(\nabla \cdot B)(x) = 0$. To satisfy $(\nabla \cdot B)(x) = 0$, we require

$$\exp[iB(\nabla f)] = I \qquad (3.12)$$

From equation (3.9) and the relation $(\nabla f)^*G = G$, we see that equation (3.12) can be satisfied if and only if

$$Z_{\nabla f}(G) = 1 \qquad (3.13)$$

The multiplier law (3.11) then requires that

$$Z_h(G) = Z_{h_\perp}(G) \qquad (3.14)$$

where $h_\perp$ is the transverse part of h.

Thus, in the representation theory of the local currents the initial-value equations, which as we saw in Section 2.3 replace gauge invariance, have the following effect. The equation $\nabla \cdot E(x) = e\rho(x)$ is used to define $\rho(x)$. The action of the operator $\exp[i\rho(f)]$ can then be expressed entirely in terms of a multiplier. The equation $\nabla \cdot B = 0$ is equivalent to a *constraint* on the multipliers. This point is discussed further in the next section.

4. Discussion

We have presented a formulation of nonrelativistic quantum electrodynamics using local currents and the electromagnetic field. The formulation is local and gauge invariant. We expect that a similar formulation of relativistic quantum electrodynamics is possible, and in fact previous work[1] on the electrodynamics of charged scalar mesons supports this view. At this point we briefly discuss a number of questions raised by the results in Sections 2 and 3.

(1) First let us discuss the relationship between this approach and one which uses a local Lorentz-covariant potential. The potential $A_\mu(x)$ is defined on a Hilbert space $\mathscr{H}$. An indefinite metric must be introduced to obtain a unitary representation of the Poincaré group. Subspaces $\mathscr{H}'$

and $\mathcal{H}''$ are defined on which the metric is positive and zero, respectively. The physical Hilbert space is given by the quotient space $\mathcal{H}_{\text{phys}} = \mathcal{H}'/\mathcal{H}''$. Maxwell's equations then hold between matrix elements of states in $\mathcal{H}_{\text{phys}}$. It is our belief that by formulating the theory using the **E** and **B** fields and local currents, we are describing $\mathcal{H}_{\text{phys}}$ directly. Furthermore, the equations of motion, including Maxwell's, are to be interpreted as operator equations. As far as we can tell, these results do not violate any known theorem requiring the use of an indefinite metric.[13-15]

(2) No charge-carrying fields occur in the formulation of quantum electrodynamics under consideration here. The local currents are fields which carry zero charge, and hence an irreducible representation of the equal-time algebra of currents and electromagnetic field strengths describes a fixed charge sector. One may well ask: How can one use this formalism to describe states with different total charge, e.g., the zero charge sector, charge one sector, etc.? We believe the answer to this question is that the Hamiltonian (2.5) and (2.14) can be given a well-defined meaning in *more than one irreducible representation of the algebra,* and that these inequivalent representations correspond to different charge sectors.

An illustration of how this could come about is provided by the following simple example. The local currents can be defined, formally, as bilinear expressions in the canonical fields. A prototype of this situation occurs in 1-dimensional quantum mechanics, where x and p may be thought of as analogous to the canonical fields and the bilinears

$$S = x^2 \qquad \dot{S} = xp + px \tag{4.1}$$

may be introduced as analogues to the local currents.

The operators S and $\dot{S}$ satisfy the equal-time algebra

$$[S, \dot{S}] = 4iS \tag{4.2}$$

An irreducible representation of this algebra is given by the relations (4.1), acting on the Hilbert space $L^2(0, \infty)$.

Now consider the harmonic oscillator Hamiltonian

$$H = \tfrac{1}{2}(p^2 + x^2) - \tfrac{1}{2} \tag{4.3}$$

This Hamiltonian can be written formally in terms of S and $\dot{S}$ as

$$H = \tfrac{1}{8}(\dot{S} - i)(1/S)(\dot{S} + i) + \tfrac{1}{2}S - \text{const.} \tag{4.4}$$

However, the harmonic oscillator Hamiltonian on the half-line is not essentially self-adjoint, but rather has a one-parameter family of self-adjoint extensions corresponding to different choices of boundary conditions at $x = 0$. As a result, the formal expression (4.4) can be given a well-defined meaning in more than one way. For example, it can be expressed in factored form as either

$$H_0 = \tfrac{1}{8}[\dot{S} - i(2S - 1)]^\dagger (1/S)[\dot{S} - i(2S - 1)] \tag{4.5}$$

or

$$H_1 = \tfrac{1}{8}[\dot{S} - i(2S - 3)]^\dagger (1/S)[\dot{S} - i(2S - 3)] \tag{4.6}$$

These two expressions for the Hamiltonian are associated, respectively, with the even and odd parity sectors of the usual Hilbert space for the harmonic oscillator defined on $(-\infty, \infty)$.

Finally, we note that a similar situation is encountered in distinguishing bosons from fermions when local currents are used in the formulation of nonrelativistic quantum mechanics. In this case the local current algebra and the formal expression for the Hamiltonian are the same, but unitarily inequivalent representations of the current algebra are associated with different particle statistics,[8,12] and the Hamiltonian must be defined differently in each of the different representations.

(3) In Section 3 we pointed out that the initial-value equation $\nabla \cdot \mathbf{B} = 0$ translates into a constraint on the Gel'fand–Vilenkin multiplier $Z_h(G)$. In a theory with a magnetic charge, one also finds[2] that a constraint must be imposed on this multiplier. Thus the multipliers carry physical information, and an interesting question for investigation is what this information tells us about the representations of the group defined by equations (3.1) and (3.2).

It has been known for some time that the multipliers carry further physical information[12]: Representations of the equal-time algebra (2.3), (2.7)–(2.10) corresponding to systems of bosons or fermions are distinguished by the choice of multipliers. Recently, considerable progress has been made in systematically classifying the representations of the ρ, $\mathbf{J}$ current group with respect to statistics.[16] Thus, I think the analytical tools are at hand for investigating the question raised above.

(4). In conclusion, I want to repeat a point mentioned in the Introduction: A small-coupling analysis of the manifestly gauge-invariant form of quantum electrodynamics outlined in Section 2 could lead to an interesting version of perturbation theory, having the property that each individual diagram is gauge invariant.

ACKNOWLEDGMENT

The author would like to thank Dr. Ralph Menikoff for helpful advice on the preparation of this article.

References and Notes

1. D. H. Sharp, *Phys. Rev.* **165,** 1867 (1968).
2. R. Menikoff and D. H. Sharp, *J. Math. Phys.* **18,** 471 (1977).
3. F. Strocchi and A. S. Wightman, *J. Math. Phys.* **15,** 2198 (1974).
4. B. DeWitt, *Phys. Rev.* **125,** 2189 (1962).
5. S. Mandelstam, *Ann. Phys. (N.Y.)* **19,** 1 (1962).
6. P. Jordan and W. Pauli, *Z. Phys.* **47,** 151 (1928).
7. A brief review of the work on local currents in nonrelativistic quantum theory through 1974 can be found in D. H. Sharp, What we have learned about representing local nonrelativistic current algebras, in *Local Currents and Their Applications,* Eds. D. H. Sharp and A. S. Wightman, North-Holland, Amsterdam (1974), pp. 85–98. Further developments are described in Reference (8).
8. R. Menikoff and D. H. Sharp, *J. Math. Phys.* **16,** 2341 (1975).
9. G. A. Goldin and D. H. Sharp, Lie algebras of local currents and their representations in *Group Representations in Mathematics and Physics, Battelle Seattle 1969 Rencontres,* Ed. V. Bargmann, Springer-Verlag, New York (1970), pp. 300–311.
10. G. A. Goldin, J. Grodnik, R. T. Powers, and D. H. Sharp, *J. Math. Phys.* **15,** 88 (1974).
11. I. Gel'fand and N. Vilenkin, *Generalized Functions,* Vol. 4, Academic, New York (1964).
12. G. A. Goldin, *J. Math. Phys.* **12,** 462 (1971).
13. F. Strocchi, *Phys. Rev.* **162,** 1429 (1967).
14. F. Strocchi, *Phys. Rev. D* **2,** 2334 (1970).
15. R. Ferrari, L. Picasso, and F. Strocchi, *Commun. Math. Phys.* **35,** 25 (1974).
16. G. A. Goldin, R. Menikoff, and D. H. Sharp, Particle statistics from induced representations of a local current group, accepted for publication in *J. Math. Phys.*

17

On a New, Finite, "Charge-Field" Formulation of Classical Electrodynamics

Darryl Leiter

The "charge-field" approach to electrodynamic processes is based on the paradigm that charges and their associated electromagnetic fields are permanently connected in elementary charge–field functional structures, with physical processes being described by the interactions between various charge–field entities in the system. The new formulation of interest to us here, called classical elementary measurement electrodynamics (CEMED),[1] resembles Maxwell–Lorentz theory in that the fields connected to each charge are not eliminated, but differs in that each charge carries its *own* Maxwell field equation (i.e., N charges and N Maxwell "charge-fields," *each* with its *own* Maxwell field equation, and requiring its *own* independent set of boundary conditions). It resembles a direct-action theory in that only interparticle interactions play a role in the formalism, with no self-Coulomb interactions at the classical level, and that all free fields uncoupled to charges are absent from the theory. However, it differs from standard Fokker-type direct-action theories[2] in that the coupled fields are still present, i.e., not eliminated *a priori* as dynamical variables. The reason this can be done is that the Maxwell field equations (one for each charged particle) are treated like

Darryl Leiter • Code 660, NASA/Goddard Space Flight Center, Greenbelt, Maryland 20771.

identities which prescribe how fields are functionals of the currents (i.e., the Maxwell equations are treated like constraints on the particle dynamics, with free fields playing no role in the formalism since they don't represent an interaction between charges). This direct (charge–field) action formalism with the requirements of positive definite energy and causality[3] yields a theory which is in complete agreement with that of renormalized Maxwell–Lorentz theory but without any infinite renormalizations being required.

In this paper we will develop the most general version of the charge-field formalism by first formulating the theory in its most elegant form (which assumes the Maxwell charge-field to be an $N \times N$ matrix) for a theory of N classical electrons in interaction. The absence of the infinite Coulombic self-energy is shown to be guaranteed by the dynamic structure of the charge–field interaction occurring in the Lagrangian.

Consider N point charges whose trajectory dynamic variables are represented in the matrix form ($N \geq 2$)

$$
x_\mu = \begin{pmatrix} x_\mu^{(1)} & & & & 0 \\ & x_\mu^{(2)} & & & \\ & & \cdot & & \\ & & & \cdot & \\ & & & & \cdot \\ 0 & & & & x_\mu^{(N)} \end{pmatrix} \tag{1}
$$

Associated with this matrix will be a current-density matrix,

$$
J_\mu = \begin{pmatrix} J_\mu^{(1)} & & & & 0 \\ & J_\mu^{(2)} & & & \\ & & \cdot & & \\ & & & \cdot & \\ & & & & \cdot \\ 0 & & & & J_\mu^{(N)} \end{pmatrix} \qquad J_\mu^{(k)} = \frac{q(k)}{c} \frac{dx_\mu^{(k)}}{dt} \delta^3(\mathbf{x} - \mathbf{x}^{(k)}(t)) \tag{2}
$$

a mass matrix,

$$
M = \begin{pmatrix} m(1) & & & & 0 \\ & m(2) & & & \\ & & \cdot & & \\ & & & \cdot & \\ & & & & \cdot \\ 0 & & & & m(N) \end{pmatrix} \tag{3}
$$

a proper-time matrix and a four-velocity matrix given by

$$d\tau = \gamma^{-1}dt \qquad \gamma = \begin{pmatrix} \gamma(1) & & & 0 \\ & \gamma(2) & & \\ & & \cdot & \\ & & & \cdot & \\ 0 & & & & \gamma(N) \end{pmatrix} \qquad (4)$$

$$\gamma(k) = \frac{1}{[1 - (v^{(k)}/c)^2]^{-1/2}}$$

$$U^{\mu} = d\tau^{-1}(dx^{\mu}) = \gamma(dx^{\mu}/dt) \qquad (5)$$

and finally the N Maxwell charge–field degrees of freedom, represented in terms of the matrix forms

$$A_{\mu} = \begin{pmatrix} A_{\mu}^{(1)} & & & 0 \\ & A_{\mu}^{(2)} & & \\ & & \cdot & \\ & & & \cdot & \\ 0 & & & & A_{\mu}^{(N)} \end{pmatrix} \qquad (6)$$

$$F_{\mu\nu} = A_{\mu,\nu} - A_{\nu,\mu} \qquad \bar{F}_{\mu\nu} = \tfrac{1}{2}\epsilon_{\mu\nu\alpha\beta}F^{\alpha\beta} \qquad (7)$$

The action integral which gives the finite formulation of the charge–field electrodynamics in this matrix language is

$$I = \int d\tau \, \mathrm{Tr}(M(U_{\mu}U^{\mu})^{1/2}) \qquad (8)$$
$$+ \int dx^4 \{ \tfrac{1}{4}[\mathrm{Tr}(F_{\mu\nu})\mathrm{Tr}(F^{\mu\nu}) - \mathrm{Tr}(F_{\mu\nu}F^{\mu\nu})]$$
$$+ [\mathrm{Tr}(J_{\mu})\mathrm{Tr}(A^{\mu}) - \mathrm{Tr}(J_{\mu}A^{\mu})]\}$$

It is explicitly seen that *in (8) the self-interaction terms* $\mathrm{Tr}(F_{\mu\nu}F^{\mu\nu})$ and $\mathrm{Tr}(J_{\mu}A^{\mu})$ (which will be associated with the time-symmetric self-Coulomb interaction) *are dynamically excluded from this action* and hence can be expected to never reappear to create self-energy infinities in the equations of motion. We also note that, *before the variation,* the point-charge current matrix in equation (2) obeys the current conservation identity

$$J_{\mu},^{\mu} \equiv 0 \qquad (9)$$

To obtain the equations of motion from (8) we vary the matrix dynamical variables δx^μ and δA^μ, respectively, to obtain the matrix equations

$$M(du^\mu/dt) = \int dx^3 \, [\text{Tr}(F_\nu{}^\mu) - F_\nu{}^\mu]J^\nu \tag{10}$$

and

$$(\text{Tr}(F_{\mu\nu}) - F_{\mu\nu})^{,\nu} = (\text{Tr}(J_\mu) - J_\mu) \tag{11}$$

Taking the matrix trace of equation (11) yields

$$(N - 1)\text{Tr}(F_{\mu\nu}{}^{,\nu}) = (N - 1)\text{Tr}(J_\mu) \tag{12}$$

Now since $N \geqq 2$ (in order for the charge–field formalism to contain interactions) then $(N - 1) \neq 0$ and (11), (12), and (7) yield

$$F_{\mu\nu}{}^{,\nu} = J_\mu \qquad \bar{F}_{\mu\nu}{}^{,\nu} \equiv 0 \tag{13}$$

Although this equation looks formally like the usual set of Maxwell's equations, it is actually an $N \otimes N$ matrix equation associated with the particle equation of motion (10). *It is this matrix property* (associated with the charge–field concept that each charged particle carries its own Maxwell charge–field as part of its dynamic structure before the variation is performed) *which allows (10) and (13) to yield a finite formulation* of classical electron theory without infinities occurring. We can see this as follows: First we note that the conserved electromagnetic energy–momentum tensor obtained from (10) and (11) is

$$T_{\mu\nu} = \left(g_{\mu\nu}\tfrac{1}{4}[\text{Tr}(F_{\alpha\beta})\text{Tr}(F^{\alpha\beta}) - \text{Tr}(F_{\alpha\beta}F^{\alpha\beta})] \right. \tag{14}$$
$$\left. + \{\text{Tr}(F_{\mu\alpha})\text{Tr}(F^\alpha{}_\nu) - \text{Tr}(F_{\mu\alpha}F^\alpha{}_\nu)\}\right)$$

and that it obeys the conservation law

$$T_{\mu\nu}{}^{,\nu} = \text{Tr}(J_\nu)\text{Tr}(F_\mu{}^\nu) - \text{Tr}(J_\nu F_\mu{}^\nu) \tag{15}$$

From (15) the "electromagnetic" energy associated with this formalism is of the form

$$\mathscr{E}_{(em)} = \int dx^3 \, T_{00} = \int dx^3 \tfrac{1}{2}[(\text{Tr}\,\mathbf{E})^2 + (\text{Tr}\,\mathbf{B})^2 - \text{Tr}(\mathbf{E}^2 + \mathbf{B}^2)] \tag{16}$$

which is not automatically positive definite, as in the conventional Maxwell–Lorentz formulation. However, we can insure that (16) has a positive-definite value by imposing the necessary boundary conditions on (13). The most general "charge–field" solution to (13) is a mixture of time-symmetric and time-antisymmetric charge–field potentials $a_{\mu_{(\pm)}}(x)$, where in the Lorentz gauge $a_{\mu_{(\pm)}}{}^{,\mu} = 0$, and

$$a_{\mu_{(\pm)}}(x) \equiv \int dx'^4 \, D_{(\pm)}(x - x')J_\mu(x') \tag{17}$$

A simple formal solution to (13) can be written as

$$A_\mu = [a_{\mu_{(+)}} + \beta a_{\mu_{(-)}} + \text{Tr}(\eta a_{\mu_{(-1)}})I] \tag{18}$$

where β and η determine the degree of mixing of the time-symmetric and time-antisymmetric "charge-fields" occurring in (16). It can easily be checked that the boundary condition associated with (16) having a positive-definite "retarded" value in the case of like charges is

$$\beta = 0 \qquad \eta = (I/N - 1) \qquad (N \geq 2) \tag{19}$$

For this choice of boundary conditions on the charge-field solutions to (13), the solutions can be written as

$$A_\mu = a_{\mu_{(+)}} + \left(\text{Tr}(a_{\mu_{(-1)}})/(N - 1)\right)I \tag{20}$$

and, when inserted into (16), yield the positive-definite form

$$\mathscr{E}_{em} = \mathscr{E}(\text{finite, retarded Maxwell–Lorentz energy})$$
$$+ \mathscr{E}(\text{total coupled radiation charge-field energy}) \tag{21}$$

where

$$\mathscr{E}(\text{finite, retarded Maxwell–Lorentz energy}) =$$
$$\int dx^3 \tfrac{1}{2}[(\text{Tr}e_{\text{ret}})^2 + \text{Tr}(b_{\text{ret}})^2 - \text{Tr}(e_{(+)}^2 + b_{(+)}^2)] \tag{22}$$

is the *already finite remainder of the "renormalized" Maxwell–Lorentz energy* (occurring automatically via the dynamics of the formalism) and

$$\mathscr{E}(\text{total coupled radiation charge-field energy}) =$$
$$\int dx^3 \frac{(\text{Tr}(e_{(-)}))^2 + (\text{Tr}(b_{(-)}))^2}{2(N - 1)} \tag{23}$$

is the "total coupled radiation field" of the system, where $e_{(\underset{\text{ret}}{\pm})}$ and $b_{(\underset{\text{ret}}{\pm})}$ are calculated from $f_{\mu\nu(\underset{\text{ret}}{\pm})} = (a_{\mu,\nu} - a_{\nu,\mu})(\underset{\text{ret}}{\pm})$ and $D_{(\pm)}(\mathbf{x}) \equiv \tfrac{1}{2}(D_{\text{ret}}(\mathbf{x}) \pm D_{\text{adv}}(\mathbf{x})$. *This new feature, the* "total coupled radiation field," takes the role *that the "free" radiation fields play in the Maxwell–Lorentz theory.* Now having chosen boundary conditions on the solutions to (13), so that the electromagnetic energy (16) is positive definite for like charges, we substitute (20) into the equations of motion for charged particles (10) to obtain the interacting charge-field:

$$\text{Tr}(F_{\mu\nu}) - F_{\mu\nu} = \text{Tr}(F_{\mu\nu})_{(\text{ret})} - F_{\mu\nu_{(\text{ret})}} + F_{\mu\nu_{(-)}} \tag{24}$$

The particle equation of motion then obtained by equating matrix components in equations (10) and (11), is given as

$$m(k)(du^{\mu^{(k)}}/dt) = \frac{q(k)}{c} \frac{dx^{\nu^{(k)}}}{dt} \left[\left(\sum_{\substack{J \neq k \\ J=1}}^{N} f^{\mu(J)}_{\nu_{(ret)}} \right) + f^{\mu(k)}_{\nu_{(-)}} \right] \qquad (25)$$

which is the finite, physical Lorentz–Dirac equation for classical electrons. *This has been obtained without any infinite renormalization steps being used and resolves in an elegant fashion the problem of infinities in microscopic classical electron theory.* It has also been shown[4] that a macroscopically averaged charge–field of the form

$$\langle F_{\mu\nu} \rangle = \left\langle \frac{1}{N-1} \sum_{k=1}^{N} \left(\sum_{\substack{J \neq k \\ J=1}}^{N} f^{(J)}_{\mu\nu_{(ret)}} + f^{(k)}_{\mu\nu_{(-)}} \right) \right\rangle_{\text{space average}} \qquad (26)$$

is a field which can be associated with the usual macroscopic Maxwell–Lorentz electrodynamics, as an approximation in the macrocosm. However, we now see that Maxwell–Lorentz electrodynamics is not an exact formalism for the microcosm, since it can be derived as an approximation to the microscopic charge–field formalism in the macrocosm. This suggests that the inherent self-energy infinities in Maxwell–Lorentz theory, which appear in the microscopic application of it, are not fundamental. Rather they are induced by the application of an essentially macroscopic approximation to a microscopic domain. If this is true, it may also be that the self-energy infinities in usual quantum electrodynamics are not fundamental for the same reason, since the formalism is essentially a second-quantized version of Maxwell–Lorentz theory. The real test of this assertion is the generalization of the charge–field theory to the second-quantized domain. Since the electromagnetic fields are always coupled to their charges, then in a Dirac "hole-theoretic" formalism, in which the charges are second quantized with Dirac anticommutation relations, the charge–fields automatically become second quantized, since they are functionals of the current operators. Preliminary investigations have suggested that "charge–field" photon operators (which can create and annihilate quanta) can be generated this way.[5] Further research is being done to clarify the exact form of a self-consistent second-quantized charge–field formulation of quantum electrodynamics. *The existence of a finite and self-consistent classical formulation serves as a strong impetus for further investigation into the quantum domain.*

References and Notes

1. Darryl Leiter, *Ann. of Phys. (N.Y.)* **51**, 561 (1969); *Nuovo Cimento B* **23**, 391 (1974).
2. For a particularly lucid discussion of Fokker-type theories see W. Panofsky and M.

Phillips, *Classical Electrodynamics,* 2nd ed., Addison-Wesley, Reading, Massachusetts (1962).

3. By "causality" we will specifically mean the presence of "mutual" retarded electromagnetic charge-fields (see the second reference quoted in footnote 1 above); however, charge-field "radiation" can still contain "advanced" components.

4. L. Clifford, *Am. J. Phys.* **43,** 551 (1975).

5. Darryl Leiter, *Nuovo Cimento B* **48,** 15 (1978); On the quantum electrodynamics of mutually interacting charge-fields: A new approach to photon quantization in electron-positron processes.

18

Analytic Coherent States for Generalized Potentials

Michael Martin Nieto and L. M. Simmons, Jr.

1. Introduction

This is work[1,2] which I (MMN) and Mike Simmons have been doing at Los Alamos.† It's hopefully going to be short, sweet, and to the point. I have divided the talk into sections. In Section 2 I'm going to review in some detail the properties of the coherent states for the harmonic oscillator. Then (in Section 3) I'm going to describe what I will call the "classical motion generalization." By that I mean I will define "coherent states" not only for the simple harmonic oscillator but for particles in different potentials. These coherent states should follow the classical motion of a particle in such a classical potential. Finally, to show that our proposed generalization is indeed a good one, I am going to present in Section 4 a specific example which we have analytically beaten to death, and then close with a discussion.

2. Harmonic-Oscillator Coherent States

You probably all are familiar with the simple harmonic-oscillator problem. It is one of the first you see in your beginning course on

---

† This written form of my presentation is modified from a tape recording of the talk which was actually given at the symposium. In places I will insert further comments, sometimes concerning results which have been obtained since then. These will be surrounded by square brackets. Questions and answers that arose during and after the talk are included.

Michael Martin Nieto and L. M. Simmons, Jr. • Theoretical Division, Los Alamos Scientific Laboratory, University of California, Los Alamos, New Mexico 87545. Work supported by the United States Department of Energy.

quantum mechanics. The eigenstates ψ_n are the solutions to the eigen-value equation

$$H\psi_n = E_n\psi_n = \left[-\frac{\hbar^2}{2m}\frac{d^2}{dx^2} + \tfrac{1}{2}m\omega^2x^2 \right]\psi_n \qquad (2.1)$$

They are given in terms of Hermite polynominals,

$$\psi_n = (a_0/\pi^{1/2}2^n n!)^{1/2}\exp(-\tfrac{1}{2}a_0^2x^2)H_n(a_0x) \qquad (2.2)$$

$$a_0 \equiv (m\omega/\hbar)^{1/2} \qquad (2.3)$$

and have eigenenergies

$$E_n = \hbar\omega(n + \tfrac{1}{2}) \qquad (2.4)$$

In particular, and this will be important later on, note that for the simple harmonic oscillator the difference in energy between adjacent eigenstates is independent of n. It is always the constant $\hbar\omega$ no matter what n.

As you all probably also know, the position and momentum can be defined as the sum and difference of the lowering and raising operators, a^- and a^+,

$$x = (\hbar/2m\omega)^{1/2}(a^- + a^+) \qquad (2.5)$$

$$p = \left(\frac{m\hbar\omega}{2}\right)^{1/2}\frac{1}{i}(a^- - a^+) \qquad (2.6)$$

The lowering and raising operators satisfy the standard commutation relation,

$$[a^-, a^+] = 1 \qquad (2.7)$$

and have the property

$$a^{\mp}|n\rangle = (n + \tfrac{1}{2} \mp \tfrac{1}{2})^{1/2}|n \mp 1\rangle \qquad (2.8)$$

Equation (2.8) tells us that a^- and a^+ are also independent of n.

Given the above, the "coherent states" for the simple harmonic oscillator can be defined in many equivalent ways. Although these ways are equivalent for the harmonic oscillator it is not clear that they will be equivalent in general. [Indeed, they are not.]

2.1. Minimum-Uncertainty Coherent States (MUCS)

One definition of the coherent states comes from the uncertainty relation between x and p. Since

$$[x, p] = i\hbar \qquad (2.9)$$

one has

$$\Delta x \Delta p = \hbar/2 \qquad (2.10)$$

The coherent states are those states which satisfy the equality in (2.10), *subject* to the condition that

$$\Delta x / \Delta p = 1/m\omega \qquad (2.11)$$

The point in (2.11) is commonly fluffed over. All minimum-uncertainty states are not coherent states for all oscillators. A subset of them, restricted by (2.11), are coherent states for a particular oscillator. (We'll see that very clearly in a minute or two.) Note that with the restriction (2.11), the ground state is one of the coherent states.

Anyway, this is one definition of the coherent states. The definition yields that the coherent states are particular Gaussians. How do you get that? This comes from the general result that if you have three Hermitian operators A, B, and C satisfying

$$[A, B] = iC \qquad (2.12)$$

(in general C is a q-number not a c-number), this implies that

$$(\Delta A)^2 (\Delta B)^2 \geq \tfrac{1}{4}\langle C \rangle^2 \qquad (2.13)$$

The states which will satisfy the equality in this generalized uncertainty relation are given by the solutions to the eigenvalue equation

$$\left[A + \frac{i\langle C \rangle}{2(\Delta B)^2} B \right] \psi = \left[\langle A \rangle + \frac{i\langle C \rangle}{2(\Delta B)^2} \langle B \rangle \right] \psi$$

For the harmonic oscillator, with A and B being x and p, the solution to equation (2.14), subject to the restriction (2.11), is

$$\psi = [2\pi (\Delta x)^2]^{-1/4} \exp\left\{ -\left(\frac{x - \langle x \rangle}{2\Delta x} \right)^2 + \frac{i}{\hbar} \langle p \rangle x \right\} \qquad (2.15)$$

That gives the first definition of the coherent states.

2.2. Annihilation Operator Coherent States (AOCS)

Another definition of the coherent states is that they are the eigenstates $|\alpha\rangle$ of the annihilation operator a^-:

$$a|\alpha\rangle = \tfrac{1}{2}\left(\frac{x}{\Delta x} + i\frac{p}{\Delta p} \right)|\alpha\rangle = \alpha|\alpha\rangle = \tfrac{1}{2}\left(\frac{\langle x \rangle}{\Delta x} + i\frac{\langle p \rangle}{\Delta p} \right)|\alpha\rangle \qquad (2.16)$$

As is well known, these turn out to be the following linear combination of the number states:

$$|\alpha\rangle = \exp(-\tfrac{1}{2}|\alpha|^2) \sum_{n=0}^{\infty} \frac{\alpha^n}{(n!)^{1/2}} |n\rangle \tag{2.17}$$

Further, by using the generating function of the Hermite polynominals, a little algebra allows one to show that, up to a phase, equation (2.17) is the same as equation (2.15).

It is from this point of view that one can see the property (2.11) that I emphasized before. Notice that the coherent states we are discussing are in principle defined in terms of four parameters: $\langle x \rangle$, $\langle x^2 \rangle$, $\langle p \rangle$, $\langle p^2 \rangle$. But α has only two parameters, (Re α) and (Im α). You've got to eliminate two. They are eliminated by the equality in (2.10) and the restriction (2.11).

If you were to put in a different value of $\Delta x/\Delta p$ and then decompose these other states into the number states, they would not be Hermite polynomials of $a_0 x$, where a_0 is the proper a_0, but of a different $a_0' x$.

2.3. Displacement Operator Coherent States (DOCS)

Finally, there is a third standard definition. This is that one applies the unitary operator

$$A(\alpha) = \exp[\alpha a^+ - \alpha^* a^-] \tag{2.18}$$

onto the ground state,

$$|\alpha\rangle = A(\alpha)|0\rangle \tag{2.19}$$

where α is the α we defined before.

All three of the above definitions are equivalent in the simple harmonic-oscillator system. You can use any of them and you come up with the same set of states. These coherent states satisfy the following property: they will follow the classical motion no matter how small $\langle \alpha|H|\alpha \rangle = \bar{E}$. What do I mean by that? If I take a coherent state whose energy is $\bar{E}$, then $\langle \alpha|x(t)|\alpha \rangle$ will have the same motion as $x(t)$ for a classical particle of energy $(\bar{E} - E_0)$, where E_0 is the ground state energy $\hbar\omega/2$. Further, there will be no dispersion of the wave packet with time.

This concept, that these states follow the classical motion, is going to be the motivation for our generalization.

Before proceeding I wish to point out that Barut and Girardello[3] and Perelomov[4] have separately proposed generalizations of the AOCS and the DOCS which have been applicable, using group theory methods, to systems with equally spaced levels, such as spin and angular momentum, including with magnetic fields.

But our question was the following: How do you apply these techniques to systems which aren't equally spaced? You usually don't have equally spaced levels. You can have a continuum. What if the system is not exactly solvable—you may not know a dynamical group or anything about it.

Finally, there was one more thing which really tickled us. Some people who are doing laser-molecular physics problems have a great interest in classical analyses of these problems, sometimes simply because that's the only game that one knows how to play. It turns out, at least according to the adherents of this school, that often these classical analyses work surprisingly well. This too made us interested in coherent states which would follow the motion of a classical particle.

So, the question we have been asking is, "Can one define states which follow the classical motion in arbitrary potentials; real honest-to-God ones?" That being said I'm not going to start with MUCS, AOCS, or DOCS. Rather, I am going to look at the classical motion as Schrödinger[5] originally did, and then see if there are any connections to MUCS, AOCS, or DOCS. We will see that the most direct connection is to the MUCS.

3. Classical Motion Generalization

OK, back to classical physics. These next equations you saw even before your first course in quantum mechanics:

$$T + V = E \tag{3.1}$$

$$\tfrac{1}{2}m\dot{x}^2 + V(x) = E \tag{3.2}$$

$$\dot{x} = (2/m)^{1/2}[E - V(x)]^{1/2} \tag{3.3}$$

Recall that for the simple harmonic oscillator with

$$V(x) = \tfrac{1}{2}kx^2 = \tfrac{1}{2}m\omega^2 x^2 \tag{3.4}$$

the solution for $x(t)$ is

$$x(t) = (2E/m\omega^2)^{1/2} \sin \omega t \tag{3.5}$$

so that

$$p(t) = m\dot{x} = m(2E/m)^{1/2} \cos \omega t \tag{3.6}$$

Note that ω is a constant, no matter how large the amplitude—no matter what the energy. That's a property of the harmonic oscillator. In general ω is classically a function of the energy. This is the reason why people can put ωm^2 in place of k, the force constant, in the potential equation (3.4).

What we propose to do, and this is our first physical *ansatz,* is to take a classical potential and find a map which changes its Hamiltonian equation into variables $X_c(x)$ and $P_c(x, \dot{x})$ such that the Hamiltonian equation is "harmonic-oscillator-like" in these variables, for a given energy. (The subscript "c" denotes classical.) That means we want to find $X_c(x)$ and $P_c(x, \dot{x})$ whose time variations are given by

$$X_c = A_c \sin \omega_c t \tag{3.7}$$

$$P_c = B_c \cos \omega_c t \tag{3.8}$$

Equation (3.7) means that [prime denotes d/dx]

$$\dot{X}_c = \dot{x} X_c' = A_c \omega_c \cos \omega_c t \tag{3.9}$$

Thus, our P_c can be given as

$$P_c = m\dot{X}_c = p X_c' \tag{3.10}$$

By the relation $\cos x = (1 - \sin^2 x)^{1/2}$, one can combine (3.3), (3.7), and (3.9), to give X_c as the solution to the equation

$$X_c' = \left(\frac{m}{2}\right)^{1/2} \omega_c \left(\frac{A_c^2 - X_c^2(x)}{E - V(x)}\right)^{1/2} \tag{3.11}$$

Putting this all back in to the original Hamiltonian equation (3.1) you obtain

$$(1/2m)P_c^2 + \tfrac{1}{2}m\omega_c^2 X^2 = (\tfrac{1}{2}m\omega_c^2 A_c^2) \tag{3.12}$$

Question (Rohrlich). Can you map the Coulomb potential into the harmonic oscillator potential?
Answer Are you talking classically or quantum mechanically?
Reply (Rohrlich). Classically.
Answer Classically, yes. The "natural variables" turn out to be just the variables which you use in solving the classical Kepler problem. $(1/r - me^2/L^2)$ is one of them and p_r is the other one. It turns out, however, that in this multidimensional problem one does not discuss things in terms of ω_c, but rather in terms of the angular velocity between apsidal distances. (This is one of the reasons we want to talk to Mostowski[6] after this talk, since this is our version of the problem he looked at.) But there is an X_c and there is a P_c, so the answer classically is, "Yes." The answer quantum mechanically is, "Almost." Although there is a mapping of the Coulomb-potential eigenstates into the harmonic-oscillator eigenstates, the mapping is not one-to-one. Almost, but not quite.

Anyway, if we have done the above we have solved the classical problem. Now what do we propose to do quantum mechanically? We say that one should consider the quantum operators which are the quantum analogues of these "natural" classical variables X_c and P_c. That is, consider

$$X_c \rightarrow X(x) \tag{3.13}$$

$$P_c \rightarrow P = \frac{\hbar}{2i}\left[\frac{d}{dx}X'(x) + X'(x)\frac{d}{dx}\right] \tag{3.14}$$

[Note that in (3.14) we have symmetrized $X'(x)$ and p.] Because X and P satisfy

$$[X, P] = i\hbar(X')^2 \tag{3.15}$$

they have the associated uncertainty relation

$$(\Delta X)^2(\Delta P)^2 \geq (\hbar^2/4)\langle(X')^2\rangle^2 \tag{3.16}$$

We claim that the appropriate coherent states are those states which satisfy the equality in (3.16); i.e., they satisfy

$$A^-\psi_\alpha = \tfrac{1}{2}\left(\frac{X}{\Delta X} + i\frac{P}{\Delta P}\right)\psi_\alpha = \alpha\psi_\alpha \tag{3.17}$$

subject to the restriction that the ground state is one of the set. (This, of course, is an alternative to setting a particular value to $\Delta X/\Delta P$.) A^- will turn out to be proportional to the ground-state annihilation operator.

[A comment to be made here is that in all our special cases, with our choice of "natural" classical variables, it is always possible to set one of the three parameters of the states which satisfy the equality in (3.16) equal to a specific value such that the ground state is one of the set. This is not possible if one were to take $X_c(x)$ and the momentum, which is classically conjugate to $X_c(x)$, as the variables. Thus, our choice of variables is appropriate.]

Therefore, we have a continuous, two-parameter set of states. That is the whole physics of it, right there. It is very simple. First year quantum mechanics students should be able to understand it. Now let's see what happens with this formulation.

4. The Symmetric Rosen–Morse Potential

We have examined in more or less great detail a number of specific potentials. Hopefully this number will continue to grow. What I am going to concentrate on here is the symmetric Rosen–Morse (RM) potential,[7,8]

$$V(x) = U_0 \tanh^2 ax \tag{4.1}$$

$$\equiv \mathscr{E}_0 s(s + 1) \tanh^2 z \tag{4.2}$$

$$\mathscr{E}_0 = \hbar^2 a^2/2m, \tag{4.3}$$

because it has built into it many of the possible problems. There's a continuum, the discrete eigenstates are not equally spaced, and, depending on how deep the potential is, I can have many bound states or few bound states. However, despite this, we can handle this thing analytically, as you will see. (We were also interested in this potential because of its relation to the soliton problem.)

Rosen and Morse[7] used this potential in 1932 to discuss diatomic molecules. They actually had it in an asymmetric form, which is a slight generalization of (4.1), and with the symmetric part being proportional to $(\cosh z)^{-2}$. For convenience we have added "1" to this so that the minimum of the well is at the origin. That makes it $(\tanh z)^2$.

As an aside, it's very interesting that in doing this work we have gone through many of the old quantum mechanics potentials, and it turned out that, at least to our knowledge, some of the properties of these solutions have never been obtained, such as the exact normalization constants. For example, for the Rosen–Morse potential things were originally solved in terms of hypergeometric functions, and then dropped. So, partially we have been going along completing the old QM papers.[2,8]

Now we will solve the classical problem, although, with foresight, in equations (4.1)–(4.3) we have already defined U_0 in terms of the parameters which will be useful quantum mechanically. The classical solutions for X_c and P_c are

$$X_c = \sinh z = \left(\frac{E}{U_0 - E}\right)^{1/2} \sin \omega_c t \tag{4.4}$$

$$P_c = p \cosh z = (2ma^2E)^{1/2} \cos \omega_c t \tag{4.5}$$

$$\omega_c = \left[\frac{2a^2(U_0 - E)}{m}\right]^{1/2} \tag{4.6}$$

You put the above in equation (3.11) and you find that it is satisfied. The classical equations of motion are

$$\dot{X}_c = P_c/m \tag{4.7}$$

$$\dot{P}_c = -m\omega_c^2 X_c \tag{4.8}$$

Now we are going to do our coherent-states procedure. The "natural" quantum operators are

$$X = X_c = \sinh z \tag{4.9}$$

$$P = \frac{\hbar}{2i} a^2 \left[\frac{d}{dz} \cosh z + \cosh z \frac{d}{dz}\right] \tag{4.10}$$

so that we are looking for states which satisfy the equality in the uncertainty relation

$$(\Delta X)^2(\Delta P)^2 \geq (\hbar^2/4)a^4 \langle \cosh^2 z \rangle^2 \qquad (4.11)$$

These states, complete with normalization constants, are

$$\psi_{cs} = N(C, B)\phi_{cs}(x) \qquad (4.12a)$$

$$\psi_{cs} \equiv \left[\frac{a\Gamma(B + \frac{1}{2} + iu)\Gamma(B + \frac{1}{2} - iu)}{\pi^{1/2}\Gamma(B)\Gamma(B + \frac{1}{2})} \right]^{1/2} (\cosh z)^{-B}$$

$$\times \exp[C \sin^{-1}(\tanh z)] \qquad (4.12b)$$

$$B \equiv \frac{1}{2} \frac{\langle \cosh^2 z \rangle}{(\Delta \sinh z)^2} + \frac{1}{2} \qquad (4.13)$$

$$C \equiv u + iv = B\langle \sinh z \rangle + \langle \cosh z \frac{d}{dz} \rangle \qquad (4.14)$$

Finally, if

$$B \equiv s \qquad (4.15)$$

this set of states includes the ground state. Therefore, imposing (4.15) gives us our coherent states.

Before continuing, I want to take a closer look at the quantum eigenvalue problem and show how one can get our "natural quantum operators" from the raising and lowering operators. (In fact, this is how we first stumbled on our "natural operators.") The exact eigenstates and eigenvalues are

$$\psi_n = N(n, s)\phi_n(x) \qquad (4.16a)$$

$$= \left[\frac{a(s - n)\Gamma(2s - n + 1)}{\Gamma(n + 1)} \right]^{1/2} P_s^{n-s}(\tanh z) \qquad (4.16b)$$

$$E_n = \mathscr{E}_0(2ns - n^2 + s) \qquad (4.17)$$

Now, because the eigenfunctions are in terms of associated Legendre functions, you can derive the raising and lowering operators. However, for those of you who may be familiar with the work of Infeld and Hull,[9] our "n" raising and lowering operators are not theirs. They obtain a factorization which is correct mathematically, but is not the useful one physically. What their factorization does is to raise and lower "n" and "s." Thus, essentially they raise and lower you from one potential to a different potential, changing the number state. To actually generate the number states for a specific potential from their operators would drive you out of your mind. Anyway, the raising and lowering operators are

$$A_n^{\pm} = (s - n) \sinh z \mp \cosh z \frac{d}{dz} \qquad (4.18)$$

It turns out that, in a generalization of the way x and p are defined in terms of a^- and a^+ for the harmonic oscillator [see equations (2.5) and (2.6)], X and P can be defined in terms of Hermitian sums and differences of the raising and lowering operators. I say, "Hermitian," since here and in general A_n^- is *not* the adjoint of A_n^+. Specifically,

$$X = \frac{1}{4(s - n)} \{[A_n^- + (A_n^+)^\dagger] + [A_n^+ + (A_n^-)^\dagger]\} \qquad (4.19)$$

$$P = \frac{\hbar a^2}{4i} \{[A_n^- + (A_n^+)^\dagger] - [A_n^+ + (A_n^-)^\dagger]\} \qquad (4.20)$$

The above shows in one more way how what we are doing is a generalization of the harmonic oscillator. This generalization is elucidated by showing how all our results go over to the harmonic oscillator in the specific limit

$$\lim_{HO} \equiv \lim_{s \to \infty, a \to 0, sa^2 \to m\omega/\hbar} \qquad (4.21)$$

It is straightforward, but somewhat involved, to show that in the limit (4.21) the eigenstates, eigenvalues, and coherent states for the RM potential go over to the same objects for the harmonic oscillator. For example, for the eigenstates you start with the associated Legendre functions and change them to $(\cosh z)^{n-s}$ times Gegenbauer polynomials. There is a relation between Gegenbauer polynomials and Hermite polynomials in the limit we are considering. With the above transformations you can show that the RM eigenfunctions become the harmonic-oscillator eigenfunctions, complete with normalizations.

The quantum equations of motion are

$$\dot{X} = -i\hbar^{-1}[X, H] = P/m \qquad (4.22)$$

$$\dot{P} = -a\{U_0 - H - \tfrac{1}{4}\mathscr{E}_0, X\} \qquad (4.23)$$

Believe it or not, the above allow one to *exactly* calculate the operators $X(t)$ and $X^2(t)$, and similarly $P(t)$ and $P^2(t)$. Specifically,

$$X(t) = e^{iHt/\hbar} X e^{-iHt/\hbar} \qquad (4.24)$$

$$= X e^{-i\omega_0 t} \left[\cos \omega_H t + i \left(\frac{\mathscr{E}_0}{U_0 - H} \right)^{1/2} \sin \omega_H t \right]$$

$$+ P e^{-i\omega_0 t} \left[\frac{1}{2ma^2(U_0 - H)} \right]^{1/2} \sin \omega_H t \qquad (4.25)$$

$$X^2(t) = \left[X^2 - \frac{1}{2}\left(\frac{H + \mathscr{E}_0}{U_0 - H - \mathscr{E}_0} \right) \right] \exp[-i4\omega_0 t]\cos(2\omega_H t)$$

$$- i\left[XQ + \frac{1}{2}\left(\frac{U_0}{U_0 - H - \mathscr{E}_0} \right) \right]\left(\frac{\mathscr{E}_0}{U_0 - H} \right)^{1/2} \exp[-i4\omega_0 t]$$

$$\times \sin(2\omega_H t) + \frac{1}{2}\left(\frac{H + \mathscr{E}_0}{U_0 - H - \mathscr{E}_0} \right) \tag{4.26}$$

where

$$\omega_0 \equiv \mathscr{E}_0/\hbar \tag{4.27}$$

$$\omega_H \equiv \left[\frac{2a^2}{m}(U_0 - H) \right]^{1/2} \tag{4.28}$$

$$Q \equiv \cosh z\,\frac{d}{dz} \tag{4.29}$$

The calculation of (4.25), for example, is done by expanding $\exp(\pm iHt/\hbar)$ in the right-hand side of (4.24) as a series of nested commutators, and then using (4.22) and (4.23) to obtain two three-term recursion relations. They can be solved. [See Reference 2 for a description.] I have written the results with X and P (or Q for convenience) on the left and all the functions of H on the right. That turns out to be useful in calculations. If you go through the same type of calculations for the harmonic oscillator you get simpler but similar looking messes. Observe that ω_H of equation (4.28) would be the classical frequency ω_c of equation (4.6) if E were substituted for H.

Question   It's an operator?
Answer   It's an operator. Right.

Returning to our coherent states of equations (4.12)–(4.15), we can calculate $\langle X \rangle$, $\langle X^2 \rangle$, $\langle P \rangle$, $\langle P^2 \rangle$, $\langle H \rangle$, and all the rest. They satisfy the equality in the uncertainty relation (3.16).

By decomposing the coherent states at $t = 0$ into the eigenstates, the time evolution of the coherent states can be given by

$$\Psi_{cs}(x, t) = e^{-iHt/\hbar}\psi_{cs}(x) \tag{4.30}$$

$$= \frac{N(C, s)}{a}\sum_{n=0}^{[s]} N^2(n, s)\exp\left[\frac{-iE_n t}{\hbar} \right]$$

$$\times O(n, C, s)P_s^{n-s}(\tanh z) + \text{continuum} \tag{4.31}$$

where $O(n, C, s)$ is the overlap. Further, one can calculate the overlap analytically. The n-even overlap function is

$$O(n\text{-even}, C, s) = \frac{\pi^{1/2}}{\Gamma(s + 1 + n)2^{s-n}} \frac{\Gamma(s - \tfrac{1}{2}n)\Gamma(s + \tfrac{1}{2} - \tfrac{1}{2}n)}{|\Gamma(s + \tfrac{1}{2} - \tfrac{1}{2}n - i\tfrac{1}{2}C)|^2}$$

$$\times {}_4F_3 \left[\begin{array}{c} s + \tfrac{1}{2} - \tfrac{1}{2}n,\ s + \tfrac{1}{2} - \tfrac{1}{2}n,\ s - \tfrac{1}{2}n,\ -n; \\ s + \tfrac{1}{2} - \tfrac{1}{2}n - i\tfrac{1}{2}C,\ s + \tfrac{1}{2} - \tfrac{1}{2}n + i\tfrac{1}{2}C,\ s + 1 - n; \end{array}\ 1 \right] \qquad (4.32)$$

with a similar relation for n-odd. The ${}_4F_3$ hypergeometric series terminates, since one of the upper indices is a negative integer, so the overlaps are finite sums. It is all very complicated, but in exact closed form.

5. Discussion

In the above decomposition I have not written out the continuum contribution. To see why, let us first look at the bound states. The bound-state superscript on the associated Legendre functions of (4.16) is $(n - s)$, $n < s$. But $(n - s)^2$ measures the energy between the top of the potential and the particle eigenenergy. That is to say, $|n - s|$ essentially is proportional to the magnitude of a wave vector. When you go into the continuum, n trys to get larger than s, and that amounts to introducing $\pm i$ in front of the quantity $|n - s|$ which is the superscript to the Legendre function. That is, the continuum wave functions go as $\exp(\pm ikz)$. So, they are plain waves at infinity which, as they come to the well, swoop through and go out again. So, with time these continuum contributions just go off to the sides, leaving the bound-state contributions confined.

Strictly speaking, there is always some overlap with the continuum (except for the coherent state which is the ground state). As stated, this will go off the sides with time. However, if you are reasonably deep in the well the overlap of the coherent state with the continuum will be exponentially small. This is consistent with the numerical calculations of Walker and Preston[10] for the Morse harmonic oscillator. Taking a harmonic-oscillator coherent state with $\langle n \rangle = |\alpha|^2$, they only needed on the order of $|\alpha|$ states to describe the time evolution. It is strictly true that there are vanishingly small pieces of all the higher states, but to numerical accuracy, they don't contribute.

Because one knows what $\psi_{cs}(x)$ is in terms of the eigenfunctions [just set $t = 0$ in equation (4.31)], all the operators involving $f(H)$ in equations (4.24)–(4.29) will just give you $f(E_n)$ when operating on each particular eigenstate contribution (ψ_n) to the decomposition of ψ_{cs}. That means that we can exactly calculate the bound-state portion of $\langle X(t)\rangle_{cs}$, $\langle X^2(t)\rangle_{cs}$, etc., and they can be evaluated numerically.

Given all our analyses, do our coherent states follow the classical motion? If you look at all these complicated formulas and make a first, rough, analytic approximation, you can show that the states follow the classical motion in this approximation. But you have to drop all kinds of terms to state this. The real question is how good is this approximation under various circumstances, like if the average energy is high or low compared to the well depth. The exact answer depends on the parameters of the system.

We are now programming our results to get precise numerical answers. Notice that the formulas are exact analytically, so you don't have to do any solving of differential equations. What we've got are just functions of x and t. Plot them! It's complicated but exact. A friend of mine by the name of Vincent Gutschick, who is a theoretical chemist by trade—he has many trades but that's his real one—is at this very moment crunching away on the time evolution of these coherent states.

Then we will complete this same program for other cases. The analytic calculations have been done completely for some potentials, and partially for others. [Now the one-dimensional analytic calculations are complete.] Finally, we will also numerically look at the time evolution of the coherent states for these different potentials.

So, the conclusion is that we have a physical ansatz from which we feel one can define coherent states which follow the classical motion of a particle in generalized potentials. For the example we have given, we know this is true in first analytic approximation. How well does their exact motion follow the classical motion? Come back in about one month and we hopefully can give you some answers.

And, in 29 minutes and 1 second, that's it. Thank you.

[After this talk was presented, a series of detailed numerical calculations were done on the coherent states for this and other potentials. These states were found to follow the classical motion quite well. In general, the deeper the state is in a well and, given that, the more eigenstates which have significant overlap with the coherent states, the longer the state coheres in terms of the number of classical oscillations. Each time the packet hits the wall, a piece of it goes off to the background, but the main hump still follows the classical motion. A computer generated film is available showing the time evolution of these states.[11] A detailed discussion of our general coherent states will appear elsewhere.[12]]

6. Questions†

Question. Can you give the connections of your formalism to those of Barut and Perelomov?

Answer. I can give a connection to Barut[3] faster than I can to Perelomov.[4] A connection to Barut's is that one uses the ground-state lowering operator in the eigenvalue equation (3.17). The connection to Perelomov's we are now not so sure of, although we have a couple of ideas. Our big problem in making a connection there is that the raising and lowering operators, as we have defined them, are n-dependent. One can make an *ad hoc* choice and use the ground-state lowering operator to make a connection to Barut's formalism. [And it is *ad hoc* since one picks out the A_0^- over the general A_n^-.] But because in Perelomov's formulation you have an exponentiation of operators, we're not exactly sure what's a proper choice. It may turn out that a connection can only be made if you modify or change the AOCS or DOCS formulations for generalized systems. That is possible. That is something which we have looked at, but which we have not come to a conclusion on yet. [Since the conference we have found a way to generalize the AOCS and DOCS concepts to nonequally spaced eigenvalue systems. In general they are not equivalent to what can be called the generalized MUCS that we have espoused here. This will be discussed elsewhere.[12]]

Q. (Bialynicki-Birula) You can also produce coherent states based on higher-n raising and lowering operators?

A. Yes, but you see the way I presented this talk is completely "bass-ackwards" from the way that we actually obtained our results. It turns out, and this is true for all our examples, that if you start by looking for those X_c and P_c which classically vary as the sine and cosine of $\omega_c t$, and then you change to the quantum-mechanical operators, by magic every single time the X and P can be defined as the Hermitian symmetric sum and difference of the raising and lowering operators—you know—with the n-dependence normalized out.

Q. [Asked about the relation of our analytic work to the group theory point of view.]

A. Willard Miller (Minnesota) has done a tremendous amount of work on the connection between group symmetries and the special functions of mathematical physics. As you saw, here the point basically was to find out that the wave functions are Legendre functions of tanh z. It turns out that in all our other examples you can do similar weird transformations. If you like the special functions of mathematical physics, it's a fun game.

Q. [Jaynes asked what was done about the eigenfunctions in the original Rosen–Morse paper.]

A. The solutions were given by them in terms of unnormalized hypergeometric functions.

Q. (Zwanziger) Can you state again the class of potentials that you've considered?

A. So far we have considered the radial part of the Coulomb problem, the inverse cosh-squared (or the tanh-squared, as discussed here today), the Pöschl–Teller potential, which limits into the infinite square well—most people con-

† This is an edited version of a tape recording of the discussion following the talk.

sider them separately but there is a certain limit connecting them, of course we have the harmonic oscillator, and the Morse potential.

Q. (Zwanziger) What's the generality of the method, then?

A. The generality of the method is that it is based on the relation between X' and X. In principle it can be solved numerically. So, people who know how to handle differential operators on computers can hopefully take an arbitrary potential, find the eigenvalues, find the raising and lowering operators, define the classical problem and therefore numerically go through the whole procedure that we have done analytically in this example. It so happens that what we have discussed is an analytic problem to see if in an analytic problem you can define states which are "coherent" in the sense that we have defined them; i.e., they follow the classical motion and don't disperse very fast. Clearly for few-level systems, and so forth, they will quickly disperse. The question is how fast, for all situations.

Q. What about the number of dimensions? Can you do anything for higher dimensions?

A. What we feel you do there is to look as we have in the Coulomb problem.[1] You look at the radial oscillation. There ω_c is replaced in the classical problem by $\theta(t)$. That's the circular variable that you work with. People[13] have beaten us to the punch azimuthally for the angular part of the problem for spherically symmetric potentials. These states are called "intelligent spin states." If you look at their solutions they amount to using $L_\pm$ raising and lowering operators in the MUCS formalism.

Q. But is that not special because there is no precession?

A. I agree, but in general you can solve all such problems numerically. You have to follow $\theta(t)$ between successive apsidal distances. The orbit will not be closed except for the Coulomb potential and the harmonic oscillator. Other systems will have to be done approximately or numerically, but in principle they can be studied this way.

Q. (Rohrlich) Have you looked at potentials with two minima?

A. Ah, that is a very good question, Fritz. The answer is we don't know yet. Let me tell you what the problem is. Clearly, quantum mechanically you've got tunneling. Classically, if you are in one well you are going to stay in one well. What I think will end up happening is that one will have a set of states which originally will start in one well oscillating back and forth and slowly will spread out. That's what we think will happen. But we do not know yet.

Q. (Rohrlich) But you think you can handle it?

A. Hope? How's that? Yeah, we have hopes, I think that is a fair way to put it. We have hopes that we can do it. Thank you very much.

[From the results of our numerical work, we think we can now give a more definitive answer to Rohrlich's question. It appears that as a matter of principle, the time evolution of a coherent wave packet in a potential with more than one minimum is no more complicated (in principle) than for a potential with one minimum. For definiteness, consider the potential $U - 2(UR)^{1/2}x^2 + Rx^4$, with the particle confined to the right at $t = 0$.

Ordinarily, with one minimum, the coherent state starts at $t = 0$ as a minimum-uncertainty wavepacket, and then slowly disperses. When one thinks of the coherent state as being a superposition of eigenstates, the superposition is such that the nodes of all the eigenstates cancel each other except for the one hump. As time evolves, this cancellation gradually is ruined.

But it is no different for our double-minimum potential. With the double minimum one has a set of almost degenerate eigenstates (even and odd), which cancel on the left and also cancel on the right, except for one hump. As time evolves, the "tunneling" can be simply viewed as the cancellation on the left gradually getting ruined, just as it is getting ruined on the right at places different from the vicinity of the classical position.]

References and Notes

1. M. M. Nieto and L. M. Simmons, Jr., *Phys. Rev. Lett.* **41**, 207 (1978).
2. M. M. Nieto and L. M. Simmons, Jr., *Phys. Rev. A* **19**, 438 (1979).
3. A. O. Barut and L. Girardello, *Commun. Math. Phys.* **21**, 41 (1971).
4. M. Perelomov, *Commun. Math. Phys.* **26**, 222 (1972), and *Usp. Fiz. Nauk* **123**, 23 (1977), [English translation in *Sov. Phys. Usp.* **20**, 703 (1977)].
5. E. Schrödinger, *Naturwissenschaften* **14**, 664 (1926).
6. J. Mostowski, *Lett. Math. Phys.* **2**, 1 (1977).
7. N. Rosen and P. M. Morse, *Phys. Rev.* **42**, 210 (1932).
8. M. M. Nieto, *Phys. Rev. A* **17**, 1273 (1978).
9. L. Infeld and T. E. Hull, *Rev. Mod. Phys.* **23**, 21 (1951).
10. R. B. Walker and R. K. Preston, *J. Chem. Phys.* **67**, 2017 (1977).
11. V. P. Gutschick, M. M. Nieto, and F. Baker, *Time evolution of coherent states for general potentials,* movie (13 min., 16 mm, color, sound) available for $130 from Cinesound Co., 915 N. Highland Ave., Hollywood, California 90038. A review of this film is given in C. A. Nelson, *Am. J. Phys.* **47**, 755 (1979).
12. A detailed series of articles on the work started in References 1 and 2 is in progress. See M. M. Nieto and L. M. Simmons, Jr., *Phys. Rev. D* **20**, 1321, 1332, 1342 (1979); M. M. Nieto, Los Alamos preprint LA-UR-79-2101; V. P. Gutschick and M. M. Nieto, Los Alamos preprint LA-UR-79-2925; M. M. Nieto, L. M. Simmons, Jr., and V. P. Gutschick, Los Alamos preprint (in preparation).
13. C. Aragone, G. Guerri, S. Salamó, and J. L. Tani, *J. Phys. A* **7**, L149 (1974); H. Bacry, *Phys. Rev. A* **18**, 617 (1978).

Index